REDESIGNING HUMANS

Also by Gregory Stock

Metaman: The Merging of Humans and Machines into a Global Superorganism

The Book of Questions

REDESIGNING HUMANS

OUR INEVITABLE GENETIC FUTURE

Gregory Stock

HOUGHTON MIFFLIN COMPANY
BOSTON • NEW YORK 2002

Visit our Web site: www.houghtonmifflinbooks.com.

Library of Congress Cataloging-in-Publication Data

Stock, Gregory.
Redesigning humans : our inevitable genetic future / Gregory Stock.
p. cm.
Includes bibliographical references and index.
ISBN 0-618-06026-X
1. Human genetics — Moral and ethical aspects. 2. Human reproductive technology — Moral and ethical aspects. 3. Genetic engineering — Moral and ethical aspects. I. Title.
QH438.7 .S764 2002
176—dc21 2001051890

Printed in the United States of America

Book design by Robert Overholtzer

QUM 10 9 8 7 6 5 4 3 2 1

To Lori,

for the warmth and beauty
she breathes into my life

Contents

REDESIGNING HUMANS

The bravest are surely those who have the clearest vision of what is before them, glory and danger alike, and yet notwithstanding go out to meet it.

—THUCYDIDES

1

The Last Human

> God and Nature first made us what we are, and then out of our own created genius we make ourselves what we want to be . . . Let the sky and God be our limit and Eternity our measurement.
>
> — Marcus Garvey (1887–1940)

WE KNOW that *Homo sapiens* is not the final word in primate evolution, but few have yet grasped that we are on the cusp of profound biological change, poised to transcend our current form and character on a journey to destinations of new imagination.

At first glance, the very notion that we might become more than "human" seems preposterous. After all, we are still biologically identical in virtually every respect to our cave-dwelling ancestors. But this lack of change is deceptive. Never before have we had the power to manipulate human genetics to alter our biology in meaningful, predictable ways.

Bioethicists and scientists alike worry about the consequences of coming genetic technologies, but few have thought through the larger implications of the wave of new developments arriving in reproductive biology. Today *in vitro* fertilization is responsible for fewer than 1 percent of births in the United States; embryo selection numbers only in the hundreds of cases; cloning and human genetic modification still lie ahead. But give these emerging technologies a decade and they will be the cutting edge of human biological change.

These developments will write a new page in the history of life, allowing us to seize control of our evolutionary future. Our coming ability to choose our children's genes will have immense social impact and raise difficult ethical dilemmas. Biological enhancement will lead us into unexplored realms, eventually challenging our basic ideas about what it means to be human.

Some imagine we will see the perils, come to our senses, and turn away from such possibilities. But when we imagine Prometheus stealing fire from the gods, we are not incredulous or shocked by his act. It is too characteristically human. To forgo the powerful technologies that genomics and molecular biology are bringing would be as out of character for humanity as it would be to use them without concern for the dangers they pose. We will do neither. The question is no longer whether we will manipulate embryos, but when, where, and how.

We have already felt the impact of previous advances in reproductive technology. Without the broad access to birth control that we take so for granted, the populations of Italy, Japan, and Germany would not be shrinking; birth rates in the developing world would not be falling. These are major shifts, yet unlike the public response to today's high-tech developments, no impassioned voices protest birth control as an immense and dangerous experiment with our genetic future. Those opposing family planning seem more worried about the immorality of recreational sex than about human evolution.

In this book, we will examine the emerging reproductive technologies for selecting and altering human embryos. These developments, culminating in germline engineering — the manipulation of the genetics of egg or sperm (our "germinal" cells) to modify future generations — will have large consequences. Already, procedures that influence the germline are routine in labs working on fruit flies and mice, and researchers have done early procedures on nonhuman primates. Direct human germline manipulations may still be a decade or two away, but methods of choosing specific genes in an embryo are in use today to prevent disease, and sophisticated methods for making broader choices are arriving every year, bringing with them a taste of the ethical and social questions that will accompany comprehensive germline engineering.

The arrival of safe, reliable germline technology will signal the beginning of human self-design. We do not know where this development will ultimately take us, but it will transform the evolutionary process by drawing reproduction into a highly selective social process that is far more rapid and effective at spreading successful genes than traditional sexual competition and mate selection.

Human cloning has been a topic of passionate debate recently, but germline engineering and embryo selection have implications that are far more profound. When cloning becomes safe and reliable enough to use in humans — which is clearly not yet the case — it will be inherently conservative, if not extremely so. It will bring no new genetic constitutions into being, but will create genetic copies of people who already exist. The idea of a delayed identical twin is strange and unfamiliar, but not earthshattering. Most of us have met identical twins. They are very similar, yet different.

Dismissal of technology's role in humanity's genetic future is common even among biologists who use advanced technologies in their work. Perhaps the notion that we will control our evolutionary future seems too audacious. Perhaps the idea that humans might one day differ from us in fundamental ways is too disorienting. Most mass-media science fiction doesn't challenge our thinking about this either. One of the last major sci-fi movies of the second millennium was *The Phantom Menace,* George Lucas's 1999 prequel to *Star Wars.* Its vision of human biological enhancement was simple: there won't be any. Lucas reveled in special effects and fantastical life forms, but altered us not a jot. Despite reptilian sidekicks with pedestal eyes and hard-bargaining insectoids that might have escaped from a Raid commercial, the film's humans were no different from us. With the right accent and a coat and tie, the leader of the Galactic Republic might have been the president of France.

Such a vision of human continuity is reassuring. It lets us imagine a future in which we feel at home. Space pods, holographic telephones, laser pistols, and other amazing gadgets are enticing to many of us, but pondering a time when humans no longer exist is another story, one far too alien and unappealing to arouse our dramatic sympathies. We've seen too many apocalyptic images of nuclear, biological, and environmental disaster to think that the path to human extinction could be anything but horrific.

Yet the road to our eventual disappearance might be paved not by humanity's failure but by its success. Progressive self-transformation could change our descendants into something sufficiently different from our present selves to not be human in the sense we use the term now. Such an occurrence would more aptly be termed a pseudoextinction, since it would not end our lineage. Unlike the saber-toothed tiger and other large mammals that left no descendants when our ancestors drove them to extinction, *Homo sapiens* would spawn its own successors by fast-forwarding its evolution.

Some disaster, of course, might derail our technological advance, or our biology might prove too complex to rework. But our recent deciphering of the human genome (the entirety of our genetic constitution) and our massive push to unravel life's workings suggest that modification of our biology is far nearer to reality than the distant space travel we see in science fiction movies. Moreover, we are unlikely to achieve the technology to flit around the galaxy without being able to breach our own biology as well. The Human Genome Project is only a beginning.

Considering the barrage of press reports about the project, we naturally wonder how much is hype. Extravagant metaphor has not been lacking. We are deciphering the "code of codes," reading the "book of life," looking at the "holy grail of human biology." It is reminiscent of the enthusiasm that attended Neil Armstrong's 1969 walk on the moon. Humanity seemed poised to march toward the stars, but 2001 has come and gone, and there has been no sentient computer like HAL, no odyssey to the moons of Jupiter. Thirty years from now, however, I do not think we will look back at the Human Genome Project with a similar wistful disappointment. Unlike outer space, genetics is at our core, and as we learn to manipulate it, we are learning to manipulate ourselves.

Well before this new millennium's close, we will almost certainly change ourselves enough to become much more than simply human. In this book, I will explore the nature and meaning of these coming changes, place them within the larger context of our rapid progress in biology and technology, and examine the social and ethical implications of the first tentative steps we are now taking.

Many bioethicists do not share my perspective on where we are heading. They imagine that our technology might become potent enough to alter us, but that we will turn away from it and reject human enhancement. But the reshaping of human genetics and biology does not hinge on some cadre of demonic researchers hidden away in a lab in Argentina trying to pick up where Hitler left off. The coming possibilities will be the inadvertent spinoff of mainstream research that virtually everyone supports. Infertility, for example, is a source of deep pain for millions of couples. Researchers and clinicians working on *in vitro* fertilization (IVF) don't think much about future human evolution, but nonetheless are building a foundation of expertise in conceiving, handling, testing, and implanting human embryos, and this will one day be the basis for the manipulation of the human species. Already, we are seeing attempts to apply this knowledge in highly controversial ways: as premature as today's efforts to clone humans may be, they would be the flimsiest of fantasies if they could not draw on decades of work on human IVF.

Similarly, in early 2001 more than five hundred gene-therapy trials were under way or in review throughout the world. The researchers are trying to cure real people suffering from real diseases and are no more interested in the future of human evolution than the IVF researchers. But their progress toward inserting genes into adult cells will be one more piece of the foundation for manipulating human embryos.

Not everything that can be done should or will be done, of course, but once a relatively inexpensive technology becomes feasible in thousands of laboratories around the world and a sizable fraction of the population sees it as beneficial, it *will* be used.

Erewhon, the brilliant 1872 satire by Samuel Butler, contains a scene that suggests what would be needed to stop the coming reworking of human biology. Erewhon is a civilized land with archaic machines, the result of a civil war won by the "anti-machinists" five centuries before the book's story takes place. After its victory, this faction outlawed all further mechanical progress and destroyed all improvements made in the previous three centuries. They felt that to do otherwise would be suicide. "Reflect upon the extraordinary advance which machines have made during the last few hundred years," wrote their ancient leader, "and note how slowly the animal and vegetable kingdoms are advancing . . . I fear none of the existing machines; what I fear is the extraordinary rapidity at which they are becoming something very different to what they are at present . . . Though our rebellion against their infant power will cause infinite suffering . . . we must [otherwise see] ourselves gradually superseded by our own creatures until we rank no higher in comparison with them, than the beasts of the field with ourselves."

Butler would no doubt have chuckled at his own prescience had he been able to watch the special-purpose IBM computer Deep Blue defeat world chess champion Garry Kasparov in May 1997. We are at a similar juncture with our early steps toward human genetic manipulation. To "protect" ourselves from the future reworking of our biology would require more than an occasional restriction; it would demand a research blockade of molecular genetics or even a general rollback of technology. That simply won't occur, barring global bio-catastrophe and a bloody victory by today's bio-Luddites.

One irony of humanity's growing power to shape its own evolution is the identity of the architects. In 1998, I spoke at a conference on mammalian cloning in Washington, D.C., and met Ian Wilmut, the Scottish scientist whose cloning of Dolly had created such a furor the previous year. Affronted by my relative lack of concern about the eventual cloning of humans, he vehemently insisted that the idea was abhorrent and that I was irresponsible to say that it would likely occur within a decade. His anger surprised me, considering that I was only speaking about human cloning, whereas he had played a role in the breakthrough that might bring it about. Incidentally, patent attorneys at the Roslin Institute, where the work occurred, and PPL Therapeutics, which funded the work, did not overlook the importance of human applications, since claims on their patent specifically cover them.

We cannot hold ourselves apart from the biological heritage that has shaped us. What we learn from fruit flies, mice, or even a cute Dorset ewe named Dolly is relevant to us. No matter how much the scientists who perform basic research in animal genetics and reproduction may sometimes deny it, their work is a critical part of the control we will soon have over our biology. Our desire to apply the results of animal research to human medicine, after all, is what drives much of the funding of this work.

Over the past hundred years, the trajectory of the life sciences traces a clear shift from description to understanding to manipulation. At the close of the nineteenth century, describing new biological attributes or species was still a good Ph.D. project for a student. This changed during the twentieth century, and such observations became largely a means for understanding the workings of biology. That too is now changing, and in the first half of the twenty-first century, biological understanding will likely become less an end in itself than a means to manipulate biology. In one century, we have moved from observing to understanding to engineering.

Early Tinkering

The best gauge of how far we will go in manipulating our genetics and that of our children is not what we say to pollsters, but what we are doing in those areas in which we already can modify our biology. On August 2, 1998, Marco Pantani cycled along the Champs Élysées to win the eighty-fifth Tour de France, but the race's real story was the scandal over performance enhancement — which, of course, means drugs.

The banned hormone erythropoietin was at the heart of this particular chapter in the ongoing saga of athletic performance enhancement. By raising the oxygen-carrying capacity of red blood cells, the drug can boost endurance by 10 to 15 percent. Early in the race, a stash of it was found in the car of the masseur of the Italian team Festina — one of the world's best — and after an investigation the entire team was booted from the race. A few days later, more erythropoietin was found, this time in the possession of one of the handlers of the Dutch team, and several of its cyclists were kicked out. As police raids intensified, five Spanish teams and an Italian one quit in protest, leaving only fourteen of the original twenty-one teams.

The public had little sympathy for the cheaters, but a crowd of angry Festina supporters protested that their riders had been unfairly singled out, and the French minister of health insisted that doping had been going on since racing began. Two years later in a courtroom in Lille, the French sports icon Richard Virenque, five-time winner of the King of the Mountains jersey in the Tour de France, seemed to confirm as much when the president of the court asked him if he took doping products. "We don't say doping," replied Virenque. "We say we're 'preparing for the race.'"

The most obvious problem with today's performance-enhancing drugs — besides their being a way of cheating — is that they're dangerous. And when one athlete uses them, others must follow suit to stay competitive. But more than safety is at issue. The concern is what sports will be like when competitors need medical pit crews. As difficult as the problem of doping is, it will soon worsen, because such drugs will become safer, more effective, and harder to detect.

Professional sports offers a preview of the spread of enhancement technology into other arenas. Sports may carry stronger incentives to cheat, and thus push athletes toward greater health risks, but the nonsporting world is not so different. A person working two jobs feels under pressure to produce, and so does a student taking a test or someone suffering the effects of growing old. When safe, reliable metabolic and physiological enhancers exist, the public will want them, even if they are illegal. To block their use will be far more daunting than today's war on drugs. An antidrug commercial proclaiming "Dope is for dopes!" or one showing a frying egg with the caption "Your brain on drugs" would not persuade anyone to stop using a safe memory enhancer.

Aesthetic surgery is another budding field for enhancement.

When we try to improve our appearance, the personal stakes are high because our looks are always with us. Knowing that the photographs of beautiful models in magazines are airbrushed does not make us any less self-conscious if we believe we have a smile too gummy, skin too droopy, breasts too small, a nose too big, a head too bald, or any other such "defects." Surgery to correct these nonmedical problems has been growing rapidly and spreading to an ever-younger clientele. Public approval of aesthetic surgery has climbed some 50 percent in the past decade in the United States. We may not be modifying our genes yet, but we are ever more willing to resort to surgery to hold back the most obvious (and superficial) manifestations of aging, or even simply to remodel our bodies. Nor is this only for the wealthy. In 1994, when the median income in the United States was around $38,000, two thirds of the 400,000 aesthetic surgeries were performed on those with a family income under $50,000, and health insurance rarely covered the procedures. Older women who have subjected themselves to numerous face-lifts but can no longer stave off the signs of aging are not a rarity. But the tragedy is not so much that these women fight so hard to deny the years of visible decline, but that their struggle against life's natural ebb ultimately must fail. If such a decline were not inevitable, many people would eagerly embrace pharmaceutical or genetic interventions to retard aging.

The desire to triumph over our own mortality is an ancient dream, but it hardly stands alone. Whether we look at today's manipulations of our bodies by face-lifts, tattoos, pierced ears, or erythropoietin, the same message rings loud and clear: if medicine one day enables us to manipulate our biology in appealing ways, many of us will do so — even if the benefits are dubious and the risks not insignificant. To most people, the earliest adopters of these technologies will seem reckless or crazy, but are they so different from the daredevil test pilots of jet aircraft in the 1950s? Virtually by definition, early users believe that the possible gains from their bravado justify the risks. Otherwise, they would wait for flawed procedures to be discarded, for technical glitches to be worked through, for interventions to become safer and more predictable.

In truth, as long as people compete with one another for money, status, and mates, as long as they look for ways to display their worth and uniqueness, they will look for an edge for themselves and their children.

People will make mistakes with these biological manipulations. People will abuse them. People will worry about them. But as much could be said about any potent new development. No governmental body will wave some legislative wand and make advanced genetic and reproductive technologies go away, and we would be foolish to want this. Our collective challenge is not to figure out how to block these developments, but how best to realize their benefits while minimizing our risks and safeguarding our rights and freedoms. This will not be easy.

Our history is not a tale of self-restraint. Ten thousand years ago, when humans first crossed the Bering Strait to enter the Americas, they found huge herds of mammoths and other large mammals. In short order, these Clovis peoples, named for the archaeological site in New Mexico where their tools were first identified, used their skill and weaponry to drive them to extinction. This was no aberration: the arrival of humans in Australia, New Zealand, Madagascar, Hawaii, and Easter Island brought the same slaughter of wildlife. We may like to believe that primitive peoples lived in balance with nature, but when they entered new lands, they reshaped them in profound, often destructive ways. Jared Diamond, a professor of physiology at the UCLA School of Medicine and an expert on how geography and environment have affected human evolution, has tried to reconcile this typical pattern with the rare instances in which destruction did not occur. He writes that while "small, long-established egalitarian societies can evolve conservationist practices, because they've had plenty of time to get to know their local environment and to perceive their own self-interest," these practices do not occur when a people suddenly colonizes an unfamiliar environment or acquires a potent new technology.

Our technology is evolving so rapidly that by the time we begin to adjust to one development, another is already surpassing it. The answer would seem to be to slow down and devise the best course in advance, but that notion is a mirage. Change is accelerating, not slowing, and even if we could agree on what to aim for, the goal would probably be unrealistic. Complex changes are occurring across too broad a front to chart a path. The future is too opaque to foresee the eventual impacts of important new technologies, much less whole bodies of knowledge like genomics (the study of genomes). No one understood the powerful effects of the automobile or television at its inception. Few appreciated that our use of antibiotics would lead to widespread drug resistance or that improved nutrition and public health in the developing world would help bring on a population explosion. Our blindness about the consequences of new reproductive technologies is nothing new, and we will not be able to erase the uncertainty by convening an august panel to think through the issues.

No shortcut is possible. As always, we will have to earn our knowledge by using the technology and learning from the problems that arise. Given that some people will dabble in the new procedures as soon as they become even remotely accessible, our safest path is to not drive early explorations underground. What we learn about such technology while it is imperfect and likely to be used by only a small number of people may help us figure out how to manage it more wisely as it matures.

Genes and Dreams

James Watson, codiscoverer of the structure of DNA, cowinner of the Nobel Prize, and first director of the Human Genome Project, is arguably the most famous biologist of our times. The double-helical structure of DNA that he and Francis Crick described in 1953 has become the universally recognized symbol of a scientific dawn whose brightness we have barely begun to glimpse. In 1998, I was the moderator of a panel on which he sat with a half-dozen other leading molecular biologists, including Leroy Hood, the scientist who developed the first automated DNA sequencer, and French Anderson, the father of human gene therapy. The topic was human germline engineering, and the audience numbered about a thousand, mostly nonscientists. Anderson intoned about the moral distinction between human therapy and enhancement and laid out a laundry list of constraints that would have to be met before germline interventions would be acceptable. The seventy-year-old Watson sat quietly, his thinly tufted head lolled back as though he were asleep on a bus, but he was wide awake, and later shot an oblique dig, complaining about "fundamentalists from Tulsa, Oklahoma," which just happens to be where Anderson grew up. Watson summed up his own view with inimitable bluntness: "No one really has the guts to say it, but if we could make better human beings by knowing how to add genes, why shouldn't we?"

Anderson, a wiry two-time national karate champion in the over-sixty category, is unused to being attacked as a conservative. Too often he has been the point man for gene therapy, receiving death threats for his pioneering efforts in the early 1990s and for a more recent attempt to win approval for fetal gene therapy. But the landscape has shifted. When organizing this symposium, a colleague and I worried about disruptive demonstrators, and could find only an occasional article outside academia on human germline therapy. A year later, stories about "designer children" were getting major play in *Time* and *Newsweek,* and today I frequently receive e-mail from high school students doing term papers on the subject.

Watson's simple question, "If we could make better humans . . . why shouldn't we?" cuts to the heart of the controversy about human genetic enhancement. Worries about the procedure's feasibility or safety miss the point. No serious scientists advocate manipulating human genetics until such interventions are safe and reliable.

Why all the fuss, then? Opinions may differ about what risks are acceptable, but virtually every physician agrees that any procedure needs to be safe, and that any potential benefit needs to be weighed against the risks. Moreover, few prospective parents would seek even a moderately risky genetic enhancement for their child unless it was extremely beneficial, relatively safe, and unobtainable in an easier way. Actually, some critics, like Leon Kass, a well-known bioethicist at the University of Chicago who has long opposed such potential interventions, aren't worried that this technology will fail, but that it will succeed, and succeed gloriously.

Their nightmare is that safe, reliable genetic manipulations will allow people to substantively enhance their biology. They believe that the use — and misuse — of this power will tear the fabric of our society. Such angst is particularly prevalent in western Europe, where most governments take a more conservative stand on the use of genetic technologies, even banning genetically altered foods. Stefan Winter, a physician at the University of Bonn and former vice president of the European Committee for Biomedical Ethics, says, "We should never apply germline gene interventions to human beings. The breeding of mankind would be a social nightmare from which no one could escape."

Given Hitler's appalling foray into racial purification, European sensitivities are understandable, but they miss the bigger picture. The possibility of altering the genes of our prospective children is not some isolated spinoff of molecular biology but an integral part of the advancing technologies that culminate a century of progress in the biological sciences. We have spent billions to unravel our biology, not out of idle curiosity, but in the hope of bettering our lives. We are not about to turn away from this.

The coming advances will challenge our fundamental notions about the rhythms and meaning of life. Today, the "natural" setting for the vast majority of humans, especially in the economically developed world, bears no resemblance to the stomping grounds of our primitive ancestors, and nothing suggests that we will be any more hesitant about "improving" our own biology than we were about "improving" our environment. The technological powers we have hitherto used so effectively to remake our world are now potent and precise enough for us to turn them on ourselves. Breakthroughs in the matrixlike arrays called DNA chips, which may soon read thirty thousand genes at a pop; in artificial chromosomes, which now divide as stably as their naturally occurring cousins; and in bio-informatics, the use of computer-driven methodologies to decipher our genomes — all are paving the way to human genetic engineering and the beginnings of human biological design.

The birth of Dolly caused a stir not because of any real possibility of swarms of replicated humans, but because of what it signified. Anyone could see that one of the most intimate aspects of our lives — the passing of life from one generation to the next — might one day change beyond recognition. Suddenly the idea that we could hold ourselves apart and remain who we are and as we are while transforming the world around us seemed untenable.

Difficult ethical issues about our use of genetic and reproductive technologies have already begun to emerge. It is illegal in much of the world to test fetal gender for the purpose of sex selection, but the practice is commonplace. A study in Bombay reported that an astounding 7,997 out of 8,000 aborted fetuses were female, and in South Korea such abortions have become so widespread that some 65 percent of thirdborn children are boys, presumably because couples are unwilling to have yet a third girl. Nor is there any consensus among physicians about sex selection. In a recent poll, only 32 percent of doctors in the United States thought the practice should be illegal. Support for a ban ranged from 100 percent in Portugal to 22 percent in China. Although we may be uncomfortable with the idea of a woman aborting her fetus because of its gender, a culture that allows abortion at a woman's sole discretion would require a major contortion to ban this sex selection.

Clearly, these technologies will be virtually impossible to control. As long as abortion and prenatal tests are available, parents who feel strongly about the sex of their child will use these tools. Such practices are nothing new. In nineteenth-century India, the British tried to stop female infanticide among high-caste Indians and failed. Modern technology, at least in India, may merely have substituted abortion for infanticide.

Sex selection highlights an important problem that greater control over human reproduction could bring. Some practices that seem unthreatening when used by any particular individual could become very challenging if they became widespread. If almost all couples had boys, the shortage of girls would obviously be disastrous, but extreme scenarios of this sort are highly suspect because they ignore corrective forces that usually come into play.

Worry over potential sex imbalances is but one example of a general unease about embryo selection. Our choices about other aspects of our children's genetics might create social imbalances too — for example, large numbers of children who conform to the media's ideals of beauty. Such concerns multiply when we couple them with visions of a "slippery slope," whereby initial use, even if relatively innocuous, inevitably leads to ever more widespread and problematic future applications: as marijuana leads to cocaine, and social drinking to alcoholism, gender selection will lead to clusters of genetically enhanced superhumans who will dominate if not enslave us. If we accept such reasoning, the only way to avoid ultimate disaster is to avoid the route at the outset, and we clearly haven't.

The argument that we should ban cloning and human germline therapy because they would reduce genetic diversity is a good example of the misuse of extrapolations of this sort. Even the birth of a whopping one million genetically altered children a year — more than ten times the total number of IVF births during the decade following the first such procedure in 1978 — would still be less than 1/100 of the babies born worldwide each year. The technology's impact on society will be immense in many ways, but a consequential diminution of biological diversity is not worth worrying about.

To noticeably narrow the human gene pool in the decades ahead, the technology would have to be applied in a consistent fashion and used a hundred times more frequently than even the strongest enthusiasts hope for. Such widespread use could never occur unless great numbers of people embraced the technology or governments forced them to submit to it. The former could happen only if people came to view the technology as extraordinarily safe, reliable, and desirable; the latter only if our democratic institutions had already suffered assaults so grave that the loss of genetic diversity would be the least of our problems. While there are many valid philosophical, social, ethical, scientific, and religious concerns about embryo selection and the manipulation of the human germline, the loss of genetic diversity is not one of them.

Flesh and Blood

As we explore the implications of advanced reproductive technologies, we must keep in mind the larger evolutionary context of the changes now under way. At first glance, human reproduction mediated by instruments, electronics, and pharmaceuticals in a modern laboratory seems unnatural and perverted. We are flesh and blood; this is not our place. But by the same token, we should abandon our vast buzzing honeycombs of steel, fiber optics, and concrete. Manhattan and Shanghai bear no resemblance to the African veldt that bore us.

Cocooned in the new environments we have fashioned, we can easily forget our kinship to our animal ancestors, but roughly 98 percent of our gene sequences are the same as a chimpanzee's, 85 percent are the same as a mouse's, and more than 50 percent of a fruit fly's genes have human homologues. The immense differences between us and the earth's other living creatures are less a result of our genetic and physiological dissimilarities than of the massive cultural construct we inhabit. Understanding this is an important element in finding the larger meaning of our coming control of human genetics and reproduction. And if we are to understand the social construction that is the embodiment of the human enterprise and the source of its technology, we need to see its larger evolutionary context.

A momentous transition took place 700 million years ago when single cells came together to form multicellular life. All the plants and animals we see today are but variations on that single theme — multicellularity. We all share a common origin, a common biochemistry, a common genetics, which is why researchers can ferry a jellyfish gene into a rabbit to make the rabbit's skin fluoresce under ultraviolet light, or use a mammalian growth-hormone gene to make salmon grow larger.

Today we are in the midst of a second and equally momentous evolutionary transition: the human-led fusion of life into a vast network of people, crops, animals, and machines. A whir of trade and telecommunications is binding our technological and biological creations into a vast social organism of planetary dimensions. And this entity's emergent powers are expanding our individual potentials far beyond those of other primates.

This global matrix has taken form in only a few thousand years and grows ever tighter and more interconnected. The process started slowly among preliterate hunter-gatherers, but once humans learned to write, they began to accumulate knowledge outside their brains. Change began to accelerate. The storage capacity for information became essentially unlimited, even if sifting through that information on the tablets and scrolls where it resided was hard. Now, however, with the advent of the computer, the power to electronically manipulate and sort this growing body of information is speeding up to the point where such processing occurs nearly as easily as it previously did within our brains. With the amount of accessible information exploding on the Internet and elsewhere, small wonder that our technology is racing ahead.

The social organism we have created gives us not only the language, art, music, and religion that in so many ways define our humanity, but the capacity to remake our own form and character. The profound shifts in our lives and values in the past century are not some cultural fluke; they are the child of a larger transformation wrought by the diffusion of technology into virtually every aspect of our lives, by trade and instantaneous global telecommunications, and by the growing manipulation of the physical and biological worlds around us.

Critical changes, unprecedented in the long history of life, are under way. With the silicon chip we are making complex machines that rival life itself. With the space program we are moving beyond the thin planetary film that has hitherto constrained life. With our biological research we are taking control of evolution and beginning to direct it.

The coming challenges of human genetic enhancement are not going to melt away; they will intensify decade by decade as we continue to unravel our biology, our nature, and the physical universe. Humanity is moving out of its childhood and into a gawky, stumbling adolescence in which it must learn not only to acknowledge its immense new powers, but to figure out how to use them wisely. The choices we face are daunting, but putting our heads in the sand is not the solution.

Germline engineering embodies our deepest fears about today's revolution in biology. Indeed, the technology is the ultimate expression of that revolution because it may enable us to remake ourselves. But the issue of human genetic enhancement, challenging as it is, may not be the most difficult possibility we face. Recent breakthroughs in biology could not have been made without the assistance of computerized instrumentation, data analysis, and communications. Given the blistering pace of computer evolution and the Hollywood plots with skin-covered cyborgs or computer chips embedded in people's brains, we naturally wonder whether cybernetic developments that blur the line between human and machine will overshadow our coming ability to alter ourselves biologically.

The ultimate question of our era is whether the cutting edge of life is destined to shift from its present biological substrate — the carbon and other organic materials of our flesh — to that of silicon and its ilk, as proposed by leading artificial-intelligence theorists such as Hans Moravec and Ray Kurzweil. They believe that the computer will soon transcend us. To be the "last humans," in the sense that future humans will modify their biology sufficiently to differ from us in meaningful ways, seems tame compared to giving way to machines, as the Erewhonians so feared. Before we look more deeply at human biological enhancement and what it may bring, we must consider what truth these machine dreams contain.

2

Our Commitment to Our Flesh

> It is in moments of illness that we are compelled to recognize that we live not alone but chained to a creature of a different kingdom, whole worlds apart, who has no knowledge of us and by whom it is impossible to make ourselves understood: our body.
>
> — Marcel Proust, *À la Recherche du Temps Perdu*

Cyborg Fantasies

Progress in fields such as artificial intelligence, bio-informatics, and the design of computer chips will greatly influence the pace and extent of our ability to reshape human biology. The question is whether we will soon incorporate these surging technologies into ourselves in a major way.

In *Johnny Mnemonic,* the 1995 film based on William Gibson's cyberpunk story of the same name, Keanu Reeves plays a "mnemonic courier" who makes his living in 2021 by smuggling data in a "wet-wired brain implant" (apparently for clients who can't figure out how to send encrypted e-mail files). In a job that was to buy him his retirement, an oversized data file he loads through a little plug at the back of his head not only leads to frothing and jaw clenching, but also threatens fatal "synaptic seepage" within a day if he doesn't unload it, which spins him into a maelstrom of conflict.

The gritty union of biology and technology depicted in the film is an unhappy one filled with jarring, violent landscapes of wrecked cities, cyborgs, and hypertechnology run amuck, but it evokes the idea that such invention soon might lead us toward a potent yet more benign human enhancement. I think not. People may dream of enhancing their minds by embedding chips in their brains, but a sophisticated interface between our nervous system and silicon would be incredibly complex.

Hollywood images of humanlike cyborgs lull our thinking, because they so completely ignore the messy realities of basic physiology. If a detail like wound healing comes up at all in a sci-fi fantasy about human chip-heads, some unspecified advanced technology usually mends the incision in seconds. This is mere theater. The inner terrain of our brain resembles neither the neat geometry of the computer chip nor the abstract corridors of cyberspace; it is flesh and blood. Wounds heal slowly. Pain lingers. Aging skin sags.

When we're healthy, we may persuade ourselves that we aren't hostage to our body's needs, but once disease penetrates our defenses, reality quickly banishes this conceit. Biology has never been easy to control. Some fifty years ago, wonder drugs such as penicillin, streptomycin, and chloramphenicol promised a rapid end to bacterial infections, but the arrival of multidrug resistance in the 1980s showed the naiveté of that early hope. Our victory over infectious diseases was provisional. The war not only goes on; we are trying to cope with new strains of tuberculosis and other diseases for which drug resistance has become a deepening problem.

Our current inability to vanquish bacterial disease might make some people cautious about making grandiose predictions about installing chips in our brains, but Ray Kurzweil, the inventor of the Kurzweil Reading Machine, the Kurzweil music synthesizer, and other high-tech products, does not hesitate. "Porting our brains to new computational mechanisms will not happen all at once," he writes in *The Age of Spiritual Machines.* "We will enhance our brains gradually through direct connection with machine intelligence until the essence of our thinking has fully migrated to the far more capable and reliable new machinery." Such a future sounds almost plausible. Technology, after all, is evolving rapidly, and who knows how far it will eventually take us? But Kurzweil's vision is not of some *distant* future; he thinks these changes will be here within decades. He predicts that by 2029, computer technology will have progressed to the point where "direct neural pathways have been perfected for high-bandwidth connection to the human brain"; where "a range of neural implants is becoming available to enhance visual and auditory perception and interpretation, memory, and reasoning"; and where there is "widespread use of all-encompassing visual, auditory, and tactile communication using direct neural connections." To top this off, he predicts that a mere century hence, the distinction between humans and computers will no longer be clear.

Such techno-exuberance, though an increasing influence on our culture, is far-fetched. Our flesh is a dense three-dimensional matrix of biological cells, ill suited for a permanent, working union with broad arrays of sensitive electronic probes. To grasp the immense challenge of splicing electronics into our nervous system, one need only look at an electron micrograph of the brain. It shows a crowded tangle of cellular bodies and dendrites, not some neat textbook schematic. Moreover, we each have unique self-organizing connections. The idea of keeping track of millions of neurons in a nerve bundle and tapping into them like phone lines is a huge leap of faith.

Physicians have implanted electrodes in specific locations in the body — the cochlea of the ear, for example, or even the brain stem, to restore rudimentary hearing — and patients sometimes recover a significant amount of sensory function. But erratic success with a few electrodes is a far cry from the reliable scaffolding of implanted leads needed to achieve a comprehensive linkage with the brain. And we are much more forgiving in measuring success when the goal is to ameliorate profound sensory deficit rather than enhance normal functioning, which presumably is the ultimate goal of a union with computers.

We must not discount the gulf between partially repairing bodily defects and enhancing healthy functioning. In 2001, doctors finally began implanting stand-alone battery-powered mechanical hearts in people, but miraculous as the device must seem to a patient whose own damaged heart can barely keep him or her alive, a vital young athlete would find the device unappealing. Indeed, I cannot imagine any apparatus that would serve us better than our own healthy heart, which responds so perfectly to our changing activity and emotions and is so well matched to the capacities of the rest of our circulatory system. A healthy human heart represents the ideal to which any replacement must aspire, and except for a little more durability, no improvement is possible so long as we remain otherwise unchanged. Moreover, the heart is simple compared with the brain, with its ever-shifting synaptic patterns and its intricate chemical and electrical exchanges with the rest of the body.

In clinical experiments on patients with neurological problems, researchers can stimulate pleasure, create a perception of light, and trigger memories using brain electrodes, but such interventions are crude. They are more akin to whacking a person's head with a mallet than they are to the feats of cyberpunk fiction. Only a true believer could imagine that we are at the threshold of tying into our cerebral hemispheres to gain new sensory and computational powers. Even serious conceptualization of a workable network of electrodes capable of generating a sufficiently flexible, nuanced, and predictable linkage with this organ is a distant possibility, and no amount of hand-waving about the awesome future of computers can change this.

Nor is the body our natural ally in this quest. Our flesh struggles to eliminate the foreign intrusions we conceive. Given the incompatible natures of flesh and microchips, developing brain implants that enhance thinking will likely be much harder than building a superhuman intelligence that is pure computer. As Hans Moravec, the founder of Carnegie Mellon University's robotics program, points out in *Mind Children,* his seminal reflection on the implications of continued exponential growth in computing power, once we build human-equivalent computers, they will figure out how to build superhuman ones.

The positive-feedback loop — ever smarter, smaller, and faster computers assembled in ever greater numbers to bring their skills and creativity to bear on their own evolution — will move us forward until we finally come up against the engineering limits of the physical world. At present, we can only guess how distant that frontier is, but the human mind cannot be the highest summit of cognitive performance. Accepting the possibility of our eventual displacement, however, tells us nothing about the specific path that leads there, or the length or speed of the journey. Jaron Lanier, a key figure in the development of virtual reality, suspects that the biggest obstacle to transcending the human mind will not be the realization of faster circuitry, but of software that can use it effectively.

Linking Brains and Computers

The problem with Ray Kurzweil's vision goes much deeper than mere technical feasibility. Even if thirty years from now, as he predicts, we could buy a machine with the computational power of a thousand human brains for $1,000, program it effectively, and somehow shunt it into our brains, why would we? So steeped are we in the culture of artificial intelligence and special effects that at first this sounds like a ludicrous question. Such an amazing augmentation surely would transform us mere mortals into cyber-demigods. Yet when I try to think of what I might gain by having a working link between my brain and a supercomputer, I am stymied if I insist on two criteria: that the benefits could not be as easily achieved through some other, noninvasive procedure, and that the benefits must be worth the discomforts of brain surgery. I am not being a stickler here: any healthy patient will demand far more before climbing onto the operating table.

As I see it, an actual brain-computer link would bring us almost nothing that our senses — fed by tiny external devices such as miniature speakers to whisper in our ears and fiber-optic eyeglass projectors to throw images onto our retinas — could not. We learn about the world through our senses. We are wired to respond emotionally to them. This is why our immediate future will almost certainly focus on augmenting and titillating our senses, not on carving some new high-bandwidth superhighway into our brains. We have no reason to veer away from our current path of miniaturizing and refining cell phones, video displays, and other devices that feed our senses. A global-positioning-system brain implant to guide you to your destination would seem seductive only if you could not buy a miniature ear speaker to whisper you directions. Not only could you stow away this and other such gear when you wanted a break, you could upgrade without brain surgery.

Even the familiar but improbable notion of having direct memory access to the world's knowledge banks crumbles upon examination. If we already had a tiny intelligent electronic companion seated in our ear to answer our every question, we would have little incentive to undergo brain surgery. Enticing as a direct brain linkage sounds in the abstract, virtually every scheme for one has this flaw. Healthy individuals are not going to allow some cyber-surgeon to hack into their brains to bring them enrichments that are largely obtainable in other ways.

Nor is the problem simply that noninvasive enhancements will be better than surgical ones. Situations exist in which humans will give way not to robots or cyborgs, but to immobile computers. In combat, fighter pilots have to interact rapidly with many sophisticated sensors and controls, and any edge is important. If ever there were a realm where an improved interface between human and machine would be desirable, surely this is it.

Yet as computer technology evolves, the air force seems more likely to eliminate than enhance pilots. It is already developing a pilotless fighter with distinct advantages in warfare. With no cockpit for a fragile human, such a plane could be much lighter. It could perform maneuvers that generate g-forces people cannot handle. It would be easier to store, ship, and mobilize, since pilots wouldn't have to be kept trained. It would avoid the political risks of casualties. Each plane would cost only $10 million instead of a typical fighter's $125 million. Hollywood space battles led by squadrons of plucky starship pilots are flimsy fantasy. The days of the combat pilot are numbered.

The Fyborg

At first glance, the cyborg vision of human-machine integration portrayed in science fiction, from *The Six Million Dollar Man* to

The Terminator, seems a believable consequence of current trends: the use of artificial joints, eye lenses, pacemakers, and other prosthetics that compensate for bodily deficits is growing. Neurological links such as cochlear implants, artificial-limb actuators, and retinal electrodes are improving. Electronic devices are shrinking and becoming more powerful. The bandwidth of wireless connections is growing. These trends, however, will not convert us into cyborg humanoids with mechanical and electronic body enhancements. The functional cyborg or "fyborg" is already bringing us more intriguing possibilities.

The difference between a fyborg — as conceived by the artificial-intelligence theorist Alexander Chislenko — and a cyborg is one of boundaries. Cyborgization incorporates machine components into our bodies. Fyborgization fuses us functionally, rather than physically, with machines. Some cyborgization already exists, of course, since we do incorporate devices inside the envelope of our skin. But the physical boundary between our internal and external worlds has changed little except for dental fillings and the occasional prosthetic limb, heart valve, or artificial hip. The functional boundary between these domains, however, has blurred and shifted dramatically. Hearing aids, eyeglasses, clothing, and telephones, though physically outside us, are functionally part of us. They are instances of our growing fyborgization, which competes with cyborgization and largely relegates it to body repairs. Most people with a hearing problem, for instance, have a fyborgian hearing aid rather than a cyborgian cochlear implant. People might choose a fyborgian identity chip embedded in a ring rather than a cyborgian one implanted in a finger. They might have a personal digital assistant that whispers in their ear to remind them of people's names rather than an implanted memory enhancer.

Fyborgization, which allows us to remain biological without giving up what technology offers, does not lie in the future. We already are fyborgs. We would feel diminished if we were to go about naked, eat only uncooked food, or give up our car and phone. We rely on our bank deposits more than our fat deposits to help us through lean times.

It might seem that we would want to implant devices that augment body functions such as vision and hearing, but there are strong reasons not to. We can upgrade, repair, and replace fyborgian devices more easily and link them more flexibly than cyborgian ones. Surgically installing a device makes sense only if it works much better that way and rarely needs replacement. We implant pacemakers, but only because they cannot function externally.

Computer technology offers useful insights into how we may enhance ourselves in the future. Computer "legacy" systems, for example, are mainframe dinosaurs that contain a tangled web of poorly understood and documented code from previous decades. Such systems work too well to discard but are costly to maintain, and programmers are often wary of tinkering with them for fear of disrupting something.

Our bodies are not so different. They are poorly documented, highly integrated webs of interconnected systems, and when we need medical treatment, we quite rightly worry about unanticipated side effects. We are not cavalier even about proven technologies. We delay as long as we can before replacing a leaky heart valve and cling to an external prosthesis such as a cane instead of racing out for a hip replacement. Too much can go wrong.

One way of updating a legacy computer system while minimizing changes to its core is to use programs called "functional wrappers." These mediate and transform the stream of information passing between the legacy core and the outside world. The wrapper might, for example, translate an internal format into an external standard in an outbound message, then reconvert the external response back into the legacy format for internal processing. We will be at least as cautious with ourselves. To see infrared light, for instance, we will not implant infrared sensors in our eyes but use an infrared scope to shift the light to frequencies we can see.

When programmers upgrade legacy systems, they are cautious and sometimes introduce code that operates in parallel with the legacy code, in case it has hidden functions and must be restored. We will do the same by using external devices that tie into our normal sensory channels and can be disengaged if necessary. We know we can change our minds about contact lenses or hearing aids after we've used them for a while. An implant is not so easy to remove.

That a healthy person would add a brain implant to take phone calls or install stainless-steel joints to enhance strength is unrealistic. The gains are too small, the challenges too large, the risks too great, and the alternatives too numerous. If at some distant hour our technological offspring displace us, they will do so not by merging with us and gradually displacing our flesh, but by a robust, independent machine evolution that transcends us.

Kevin Warwick, a professor in the Department of Cybernetics at the University of Reading, just west of London, thinks otherwise. An early cyborg wannabe, he made the leap in 1998 when he placed his first glass-encased chip implant just beneath the skin of his upper arm. It enabled a computer to monitor his presence by radio signals as he wandered around his department. During a nine-day trial, a voice would greet him when he entered the building; his computer would sense his approach, open the door to his office, and switch on the lights.

Fusion with a computer is a powerful image with dramatic possibilities, and when a researcher penetrates his own body to enhance rather than repair himself, it evokes a long history of science fiction. But the title of Warwick's feature story in *Wired* magazine in early 2000, "Cyborg 1.0," and the issue's cover photo of him, labeled "I, Robot," simply highlight the distance between the symbolism and the reality of implants. Such theater is a mere sideshow to the accelerating fyborgization going on all around us.

Warwick's experience with his implant touched him deeply: "I believe humans will become Cyborgs and no longer be stand-alone entities . . . ," he wrote. "I actually became emotionally attached to the computer. It took me only a couple of days to feel like my implant was one with my body. Every day in the building where I work, things switched on or opened up for me — it felt as though the computer and I were working in harmony. As a scientist, I observed that the feelings I had were neither expected nor completely explainable."

But he could have had the same experience by carrying his implant in his pocket or wearing an identity badge like those used at the MIT media labs or produced by companies pioneering wearable computing devices. More than five hundred parolees of the Florida Department of Corrections, for example, already use tamper-resistant ankle bracelets that monitor their whereabouts and alert a parole officer if they go beyond a specified area. Such tracking devices are bound to appear in the home-security market too, because many people would like to be able to summon police at the push of a button, monitor their children, or keep track of a loved one.

The emotional impact Warwick describes came from having an interactive environment, not an implant, but he fairly bubbles with enthusiasm about our coming transformation into cyborgs. "Just think," he says. "Anything a computer link can help operate or interface with could be controllable via implants: airplanes, locomotives, tractors, machinery, cash registers, bank accounts, spreadsheets, word processing, and intelligent homes . . . Will you need to learn any math if you can call up a computer merely by your thoughts? Must you remember anything at all when you can access a world Internet memory bank? . . . In the future, we won't need to code thoughts into language — we will uniformly send symbols and ideas and concepts without speaking. We will still fall back on speech in order to communicate with our newborns, however, since it will take a few years before they can safely get implants of their own, but in the future, speech will be what baby talk is today."

This is unlikely. Calculators already are great at doing arithmetic for us. If we want to free our hands from the steering wheel, we will probably hail a taxi or use some sort of programmable vehicle. As for the demise of speech, language is a rich and nuanced way of communicating thoughts and ideas. Think of all the subtleties we indicate through our choice of words, our tone of voice, our phrasings and imitations of accents and speech patterns. We have all felt the frustration of not being able to explain ourselves, but our brains have evolved so innate a capacity for language that to imagine that some imposed array of electrodes spewing signals into our cerebral cortex will do better seems wistful fantasy.

Warwick's techno-ebullience reminds me of discussions of "the house of the future" that surface every few years. The designers, hard pressed to come up with workable devices that are alluring enough to be more than conversation pieces, fill their dream houses with personalized toasters, coffee pots that adapt to our habits, and refrigerators that automatically order groceries. Devices of this sort will undoubtedly arrive, and if they are cheap and reliable, we will buy them. But such gadgetry is hard to implement well, and it will not transform our lives. The concept of ordering groceries online is appealing, but most such businesses failed because too few people thought this home delivery service worth the premium needed to cover its high costs.

Of Carbon and Silicon

Predictions of the imminent fusion of human and machine ignore the degree to which we are biological in nature and want to remain that way. Expanding our senses, enhancing our physical powers, or enlarging our minds is seductive, but until our flesh loses its vitality or becomes diseased or damaged, few of us want to replace it.

Artificial joints and pacemakers are one thing, enhancements to our central nervous system another. Film clips of a few early casualties, struggling to speak or keep their faces from twitching, would be enough to dissuade most potential users. Blood-chemistry monitors and other such devices may someday be routinely implanted, but only if they cannot be worn. Given the plethora of wristwatch-like pulse monitors and portable glucose testers for diabetics, it appears that wearable electronics is what is taking off, not implants.

Our long-term relationship with the silicon technology we are creating will likely determine humanity's future achievements, but this need not cloud our thinking about what lies ahead for us and our children. Even if the cognitive power of the computer surpasses our brainpower — as it has already done in one specialized arena after another — we ourselves will remain flesh and blood for the foreseeable future. In predicting the triumph of silicon over biology, Kurzweil and others project a doubling of computing power every year — a thousandfold rise every ten years. Carry that out until 2050, as they do, and you have more than five successive thousandfold increases in computing power. Others, however, believe that in some fifteen years, when a chip's computational elements will approach a thickness of five atoms and the leakage from quantum effects will block further shrinkage, such growth will end or greatly slow. Gordon Moore, the cofounder of Intel, who in 1965 predicted the continued doubling (known as Moore's law) that has been taking place, and even accelerating, said as much in an interview in 2000.

> The fact that materials are made of atoms is a fairly fundamental limit we have to contend with in the not too distant future. Once we get to that point, we can no longer just make things smaller. We will have to come up with other ways to increase the complexity — like making the chips bigger to get more stuff on them. I think that will change the doubling time again, from every two years to maybe four or five years. It's not the end of progress as some people have tried to paint it . . . I figure we've got another 10 or 15 years or so to keep doing what we've done in the past.

The difference between a single thousandfold increase and five such increases is immense. A stack of a thousand sheets of plastic wrap is about an inch thick. Two further thousandfold increases and that same stack is three times the height of Mount Everest. Two more and it would climb sixty times higher than the moon. Even if such growth slows, as chip elements approach atomic dimensions and designers resort to technologies like quantum and molecular computing, few doubt that progress will continue. The critical question is whether it will remain exponential, with computing power doubling with each increment of time, or if successive doublings will take longer and longer. If progress remains exponential, the doubling time doesn't make much difference in the grand scheme of things. If doubling computational power took a decade or so rather than a year or so, for example, then the transition to superhuman computing power would take a few centuries rather than a few decades. But on an evolutionary time scale, this is just quibbling; in hindsight, the transition would seem nearly instantaneous. If nonbiological complexity continues to grow exponentially, eventually it will transcend biology.

To put the timing of this potential transition in context, imagine collapsing life's 3.5-billion-year history into a period of 35 years. On this scale, some 50 years remain until the sun flares into a "red giant" and fries our planet. Our first primate ancestors appeared one year ago, thirty-four years after life began. The first hominid arrived a month ago; *Homo sapiens,* two days ago; writing, an hour ago; and the wizardry we have breathed into silicon is poised to transcend us in seconds or minutes. The timing of this possible transcendence is anybody's guess, because it depends on uncertain breakthroughs yet to come. Computing power may cease to grow exponentially or even plateau, of course, in which case biological life would remain forever king. Regardless of the ultimate outcome of the race between biological and computer evolution, however, unless the extremely rapid machine transcendence predicted by Moravec and Kurzweil plays out, *our* immediate future — involving us, our children, and our children's children — will be governed by medicine and biology.

Throughout the twenty-first century, we fyborgs will find ourselves deeply integrated into systems of machines, but we will remain biological. And as long as this is true, the primary changes to our own form and character will arise not from implants but from the direct manipulation of our genetics, our metabolism, and our biochemistry.

The continuing growth in computing power will have a huge impact on our lives, not because it will turn us into machines, but because it will allow us to create virtual realities that blur the line between the real and the imaginary, control and reshape the physical world we live in, and decipher our biology. These are the uncharted realms before us, and at their center is the coming modification of human biology, which is unique terrain, because here we are both the explorers and the explored.

The Changing Face of Biology

A recurring topic in discussions of human genetic enhancement is intelligence: how important it is, the extent to which genetics shapes it, whether we can augment it, the degree to which unequal access to technologies that boost it in children would stratify society. Interestingly, technological progress may reshape our future attitudes toward intelligence and other human attributes. A look at computational ability illustrates this. With the arrival of cheap hand-held calculators, automated cash registers, and other devices, virtually no one needs to do arithmetic mentally. I once learned a tedious method for calculating square roots. No one would bother with this now. People may worry about the dire consequences of children not learning arithmetic, but civilization has survived. Deftness at such computation impresses us, but we hardly see it as important. Just as fyborgization will inhibit cyborgization, it will stall various biological enhancements by making them less meaningful. Few parents would take any risk to bolster arithmetic skills genetically in their children, if this were possible.

Memory is similar. Today we manage such busy schedules, see so many people, and summon up information on so many topics that a good memory should be more important than ever. But it isn't. Instead, we rely on automatic dialers, electronic organizers, and other tools. As our interactions with these devices become more natural and we teach them to anticipate our needs and respond to our voices, we will find them indispensable. A good memory may one day be no more important than good eyesight is now. By the time memory enhancement becomes available, I suspect that few will be willing to risk much for it, unless it is as easy as laser eye surgery claims to be, or only means taking some "smart drug."

The social as well as the technological context is critical in determining whether a particular biological enhancement will appeal to us. As with Ray Milland's character in *The Man with the X-Ray Eyes*, who eventually rips his eyes out because he cannot stop seeing everything, having the nose of a bloodhound or the eyes of an eagle might be more than we bargain for. If our hearing or sense of smell became much more acute than other people's, the social norms that keep background noises and odors beneath our sensory thresholds would be insufficient for us. For instance, my wife's senses are very keen, and as a result she is bothered by grating sounds and obnoxious odors that I hardly notice.

It is rather poignant that we cannot yet apply technology more directly to our biological selves, because the advances in transportation, telecommunications, and other areas that enable us to transcend some of our bodily limits give us the idea that we should also be able to stay our eventual aging and decay. William Butler Yeats beautifully captures the yearning and anguish of this in "Sailing to Byzantium":

> Consume my heart away; sick with desire
> And fastened to a dying animal
> It knows not what it is; and gather me
> Into the artifice of eternity.

Nothing is more central to the trajectory and meaning of our lives than our mortality. Technological optimists like to point to the rise in life expectancy during the past century to argue that it will continue to climb in the century ahead. But most of the previous thirty-year rise resulted from public health measures, better nutrition, and the introduction of antibiotics, not from any conquest of human biology. For significant further gains in longevity, we will have to raise the maximum lifespan, which means manipulating the underlying process of aging. And some respected scientists studying the biology of aging think this is achievable. Steven Austad, a gerontologist at the University of Idaho, for example, believes that the first person who will reach the age of 150 is already alive.

Germline and other biological manipulations are a plausible way for people to overcome their bodily frailties, but a larger game is afoot. In a sense, germline manipulation is biology's bid to keep pace with the rapid evolution of computer technology. As we learn to manipulate our biology in deeper ways, the distinctions between biology and nonbiology will blur. As laboratory procedures like those now used for *in vitro* fertilization increasingly allow us to alter our genetics, the driving force of human evolution will change. Parents will select genetic modules that seem to offer the clearest benefits and the fewest risks to their future children. This will draw human reproduction under the sway of consumer marketing.

Even today, fashion and market preference determine much more than merely the selection of consumer products we find in stores. These factors also shape the biological world, determining the crops we plant, the domestic animals we raise, the flowers we grow, the pets we lavish with attention. We will not need implants to become technological creatures, because our biology itself is becoming more and more like our technology. The question is how consciously we will shape this future.

In light of the major differences we have created between poodles and Great Danes in a few thousand years, using the primitive tools of animal breeding, our own self-selection using DNA chips, artificial chromosomes, and IVF will probably change us even more, and soon.

3

Setting the Stage

We shall not cease from exploration
And the end of all our exploring
Will be to arrive where we started
And know the place for the first time.
— T. S. Eliot, "Little Gidding"

From Mainstream Research to Human Engineering

Jesse Gelsinger was feeling fine the Monday he left the sidewalk on Philadelphia's Spruce Street and passed through the modest stone archway of the Wistar Institute at the University of Pennsylvania. He had flown in from Tucson a few days earlier to participate in a phase 1 gene-therapy trial at the Institute for Human Gene Therapy there. Jesse expected no personal benefit; a phase 1 trial merely establishes safety. He wanted to help others with liver disease like his, a metabolic disorder that, without a demanding regime of tube feedings, leads to a fatal ammonia buildup in the blood.

The doctors had assured Jesse that the worst he might expect was a week of flulike symptoms, but soon after the infusion of gene-therapy adenoviruses into his liver, he developed a systemic blood-clotting disorder. Respiratory problems followed, his liver and kidneys failed, and he sank into a coma and died that Friday. The date was September 17, 1999, almost nine years to the day after French

Anderson and Michael Blaese had initiated the first officially sanctioned gene-therapy attempt by administering genetically modified T cells, critical players in our immune system, to a four-year-old girl named Ashanti DeSilva, to treat the immune disease that afflicted her.

Jesse Gelsinger's death shocked everyone. It was the first treatment-related fatality among the more than 1,100 gene-therapy patients participating in trials of adenoviral vectors, which are viruses specially modified to carry genes into certain of our cells. And as matters turned out, the Philadelphia researchers had not adequately informed Jesse of the risks. No one had told him that some previous patients in similar experiments had experienced liver damage, that a monkey had died from the procedure, or that another trial had returned adverse results.

Nor was such laxity unusual. Dr. Harold Varmus, the former director of the National Institutes of Health, wrote to Congressman Henry Waxman that researchers in similar gene-therapy trials had reported only 39 of some 691 adverse reactions to NIH's Office of Recombinant DNA Activities as required.

What a change from the heady days following DeSilva's 1990 treatment, when stories of imminent gene-therapy miracles peppered the news. Few then would have thought that after hundreds of millions of dollars of public money and enormous clinical and research efforts, we would enter the new millennium with the whole endeavor in question and not a single conclusive demonstration of effective gene therapy.

Given this background, lawmakers might easily have used the Gelsinger tragedy to crush gene-therapy research under a mountain of bureaucracy. Instead, they merely tightened oversight. The clinical possibilities remained too compelling.

Calling for caution is easy when you feel no urgency and your time is not running out. But for many people, the clock is ticking loudly. Eric Kast, a spokesman for the Cystic Fibrosis Foundation, is one of them. Not only has he no desire to slow gene-therapy research, he has read the harsh warnings of clinical-trial consent forms eight times and chosen to participate on each occasion.

"Cystic fibrosis is a genetic disease that primarily affects the lungs," he testified at the U.S. Senate hearings in 2000 investigating Gelsinger's death.

> My body produces thick, sticky mucus that clogs the airways and is difficult to cough up. Because of this, I am more susceptible to frequent lung infections, which will eventually destroy my airways. To fight these lung infections, I take antibiotics intravenously. CF also affects my digestive system, so I take pills with each meal to digest my food. When I was 31, I also developed diabetes as a result of CF, so I now take insulin shots . . .
>
> Despite this constant battle, I am fortunate. I'm still alive. I'm 33 while the current life expectancy for someone with CF is only 32 . . . Two years ago, a friend of mine with CF who was 24 was healthier than I was. Six months later, he was on a waiting list for a lung transplant, and eight months after that, he was dead. We are in a race to find a cure and we will achieve this only through strong medical research . . .
>
> Some may argue that people like me cannot agree to participate in a trial . . . because of our "desperation" to find a cure . . . I know what I am risking — my future with my family — and I know that I must take these risks to give my niece and the 30,000 others with CF the chance I never really had to live a long, healthy life.

In an ironic turn, only two months after this testimony, somatic gene therapy — which alters the genes in our soma, or body, cells and therefore does not affect the next generation — had its first clear success. Dr. Alan Fischer, at the Institut Pasteur in Paris, reported curing two infants suffering from a rare form of severe combined immunodeficiency disease by removing and genetically modifying their bone marrow cells. Later that year, a group in Montreal also reported success, this time in the treatment of hemophilia.

Such results augur well for the eventual realization of some of the promises of somatic therapy, but enormous hurdles remain. As it happens, germline interventions, which alter the genes in the first cell of the embryo and thus in every cell of a future child, may be technically easier to accomplish. Leroy Hood, the key figure in developing the automated sequencers that read the human genome, noted this at the 1998 symposium at UCLA, Engineering the Human Germline: "An amazing thing is that the manipulations to do those kinds of experiments [gene repairs] are actually much simpler in germline than in somatic therapy. If I had to project, I think fifty years from now we will be doing everything through the germline rather than in somatic tissues."

To understand why germline interventions seem easier, we must look at the fundamental challenge to all gene therapy: making a new gene active in the right place, at the right time, to the right extent. Somatic gene therapy's success relies on placing a modified gene into the cells of some target tissue and ensuring that the gene is active only there. This is no simple matter for hard-to-reach internal organs like the liver, for dispersed tissue such as muscle, and for diseases in which therapy requires the repair of nearly all afflicted cells.

Early clinical research has focused on particularly accessible tissue such as the lining of the lung and white blood cells. With cystic fibrosis, for example, researchers can use inhalants containing viruses that can carry a gene into the mucosal cells in the lung's lining. For patients with adenosine deaminase deficiency, an immune disease in which certain white blood cells are unable to function normally, doctors draw a patient's blood, extract the white cells, alter them, and infuse them back into a vein. With these therapies, the cells are relatively easy to reach, so researchers concentrate on refining the viral vectors and other methods of getting a modified gene into the cells and keeping it active.

Reaching the liver is more difficult, however, so therapists try to use the circulatory system to ferry in the viral vector or, as with Jesse Gelsinger, inject it directly using a catheter. But getting the modified virus in place is only the beginning. It also has to infect the target cells and transfer its genes so that they will function properly. And the virus has to be selective and efficient in entering the right cells. Moreover, therapists must solve these complications anew with every somatic therapy they develop.

Germline manipulations achieve selective targeting entirely differently. The need to ferry a therapeutic gene into particular tissue disappears, because that gene is already in every cell. The challenge is to regulate the gene so that it is active at the right level and at the right time and place. Compared with somatic interventions, germline insertions are in a sense more natural, in that their regulation is like that of the rest of our genome. Our skin, muscle, and brain cells, for example, have identical genes and differ only because unique subsets of these genes are active as a result of past and present signals from hormones and neighboring cells. For a newly inserted gene to operate properly, it needs a control sequence to manage its activity. We do not yet understand the manifold regulatory elements in our genome enough to design them from scratch, but we will be able to copy and imitate them.

Given the differences between somatic and germline procedures, today's sizable push to develop somatic gene therapies might seem to hold little relevance to the development of germline therapy. Somatic gene therapy, after all, is well within the accepted medical framework. No one who has watched people suffer serious illnesses like cystic fibrosis or sickle cell anemia would deny them a genetic treatment because of some vague philosophical apprehension about altering our genes.

Germline engineering represents a shift in human reproduction, but as effective somatic therapies become common, reduced public concern about genetic interventions in general will smooth the way for a move from screening and selecting embryos to actually manipulating them. In addition, research on somatic gene therapies will likely yield knowledge that can be used in germline work.

The degree of people's openness to somatic therapy is clear from Michael Blaese's gene-therapy trials on Amish and Mennonite children with Crigler-Najjar syndrome, a potentially fatal disease in which the liver enzyme that breaks down bilirubin, a product of red-blood-cell cycling, is missing. No group is more cautious than the Amish about embracing new technologies, but though they may shun television and use horses and buggies, they are welcoming the possibilities of gene therapy. In 1991, a team of Amish carpenters even gathered to build Dr. Blaese's Clinic for Special Children.

The motivation of the Amish is easy to understand. Inbreeding during the three centuries since their Swiss and German Anabaptist forebears arrived in the United States has afflicted them with a number of rare genetic diseases. Gene therapy offers Amish parents potential cures, and they — as would any parents — are grasping at the chance.

In Ira Levin's *The Boys from Brazil,* Joseph Mengele and a cabal of demented scientists in Argentina hope to restore the Third Reich by creating ninety-four clones of Hitler and arranging for each to be raised as the Führer was. The novel is gripping, but the premise that genetics could so triumph over the vagaries of personal experience and historical fortune is absurd. Individual experience is too chaotic to recreate, and the cauldron of World War I and the Great Depression that enabled Hitler to take power is gone.

The malevolent imagery that makes Levin's story riveting is not unique to cloning. The threat of malicious manipulation of the genes of human embryos makes good drama too. But the arrival of germline engineering and cloning will not require mad scientists poring over copies of *Mein Kampf.* The fundamental discoveries that spawn these coming capabilities will flow from research deeply embedded in the mainstream, research that is highly beneficial, enjoys widespread support, and certainly is not directed toward a goal like human germline engineering. The possibilities of human redesign will arrive whether or not we actively pursue them.

Activity in four overlapping areas of research — the human genome, clinical medicine, animal transgenics, and human infertility — will bring about human germline procedures. Understanding how each will contribute to germline technology provides insight into the forces that will drive our coming selection and manipulation of human embryos and the resulting shift in human reproduction.

Deciphering the Human Genome

In 1985, Robert Sinsheimer, the chancellor of the University of California at Santa Cruz, Renato Dulbecco, an Italian Nobel laureate at the Salk Institute, and Charles DeLisi, the director of the Department of Energy, independently proposed projects to sequence the 3-billion-base human genome. Only four years later, the Nobel laureate James Watson was at the helm of a $3 billion NIH effort allied with International Human Genome Organization projects in Italy, Japan, France, the United Kingdom, and other countries. Considering that the longest sequenced contiguous DNA in our genome at the time was the 67,000-base gene for human growth hormone, the goal of sequencing the entire genome in fifteen years, using as yet undeveloped technologies, was audacious.

Nonetheless, in June 2000, years ahead of schedule, Craig Venter, the CEO of Celera Genomics, and Frank Collins, the director of the Human Genome Project, announced the completion of a rough draft. With this extraordinary accomplishment, humanity took a giant step toward unraveling its biology and manipulating it in profound ways. In a sense, the race to understand and use this growing body of knowledge could start in earnest.

The immediate consequence of uncovering the more than thirty thousand human genes and their variants will be better identification of our genetic susceptibilities to various diseases and better treatments for them. At present, we use family history to ascertain our genetic risks. It is an imperfect proxy for the raw genetic information itself, but except for a few rare disorders, it is the best we have.

Often the message is not very clear. All of my uncles and aunts, for example, died of one or another cancer: one aunt died of breast cancer in her fifties, another in her sixties; one uncle (a heavy smoker) died of lung cancer in his forties, another of a brain tumor at eighty-three. My mother died of lymphatic cancer at seventy-nine; my father contracted colon cancer in his seventies, beat it, and lived into his nineties.

Of course these might represent many different diseases. But what if I could get a detailed genetic profile, identify a gene or genes whose presence indicates a predisposition to some type of cancer, and take aggressive preventive measures that might otherwise be too onerous?

The genetic tests for mutations in the BRCA1 and BRCA2 genes are an early example of such analysis. Mutations in these tumor-suppressor genes are responsible for fewer than 10 percent of breast cancers, but a woman with a family history of multiple breast cancers who tests positive has about an 80 percent risk of getting the disease. BRCA mutations are rare in the general population, with between 1 in 500 and 1 in 2,000 women having them, but even this is more common than the 1-in-3,000 frequency of mutations causing cystic fibrosis, one of the most common previously identified genetic diseases. Moreover, in the years ahead, researchers are bound to uncover additional genes that amplify or reduce the risks from the BRCA mutations.

Although the Human Genome Project offers great promise in gaining control of our evolutionary future, this is far from the minds of most scientists working on the project. They have more immediate goals: identifying disease-related genes, developing diagnostic tests, finding effective new drugs, understanding cancer and other diseases. Eric Lander, the head of the Whitehead-MIT Center for Genomic Research, the largest of the five sequencing centers for the Human Genome Project in the United States, is clear in his opposition to tinkering with our evolution. "This is the big one," he says, "the question of whether it's right to modify the genetic code so that people pass on particular traits to their children. For now, I'd like to see a ban on modifying the human germline."

Lander's caution is understandable, but it won't have any impact on the end result. Although the primary motivation for our current exploration of human biology may be largely medical, the deeper understandings needed to develop new therapies will open up the possibilities of both embryo selection and germline intervention.

Two assumptions are implicit in the idea that one day we will be able to purposefully manipulate our genes. First, genes matter and are responsible for important aspects of who we are. Second, many of the influences our genes exert are straightforward enough to identify and select or rework.

The media has hyped many recent gene discoveries, but there is no question that our genes do shape our predispositions and vulnerabilities. The complexity of most of these influences remains to be determined, but a few are surprisingly simple. It is amazing that changing a single DNA base among the 3 billion in our genome can lead to all of the complicated manifestations of an illness. Yet this is true for many diseases.

Huntington's disease involves more than one base, but its genetic underpinning is simple. If a particular triplet of "letters" — bases — in the implicated gene repeats more than thirty-five times, rather than the typical ten to twenty times, a person will end up with the disease. Huntington's strikes in midlife: a slight deterioration of the intellect, a jerking of the limbs, and a loss of balance starts an implacable ten- to twenty-five-year descent into depression, dementia, and death. The exact number of triplets gives a strong indication of both the age of onset and the rate of decline. With thirty-nine repeats, the first signs appear, on average, in the mid-sixties. With three additional repeats, onset arrives by age forty, and with another eight repeats, the afflicted will likely sink into full-blown dementia before the age of thirty.

The list of diseases resulting from single aberrant genes is long — for example, Lesch-Nyhan syndrome, which leads to mental retardation and self-mutilation, Tay-Sachs disease, which brings early neural degeneration and death, and Werner's syndrome with its symptoms of childhood aging. Each has its peculiar horrors. Disorders involving larger constellations of genes, though more difficult to unravel, will likely be even more common.

Lest we imagine, however, that every genetic aberration results in a serious medical problem, experiments called knock-out studies, in which researchers inactivate a single gene in mice or other animals to learn what it does, frequently show no noticeable consequences. This is probably because other genes with similar functions are able to "rescue" the animals. Perhaps some deficit would arise in a less pampered environment, but under laboratory conditions the altered mice sometimes seem perfectly normal.

By loose estimates, we each have a half dozen or more genetic mutations that under some conditions could cause us a problem. But most of us who are not above middle age feel reasonably healthy, don't have obvious genetic diseases, and are unaware of any genetic vulnerability that might one day undo us. Genetic technology's promise of extraordinary progress in medicine is exciting, but the impact will be diffuse for most people in their active years. A more common worry, at least in the United States, is that genetic screening might turn up something that would cost us our health insurance. Except for that, the most difficult issues we confront might not be related to health at all, but to who we are as individuals.

We will have to consider how much our genes shape personality, intelligence, athletic talent, musical ability, memory, temperament, sexual orientation. These are highly charged matters, and they have been the object of enormous debate and misinformation. Whether it is the 1994 bestseller *The Bell Curve,* in which Richard Herrnstein and Charles Murray argued that genetics is responsible for most of the racial differences in IQ test scores, or the aggressive rebuttals by Stephen Jay Gould and others, the arguments are hard to evaluate and harder still to untangle from the political and social biases of their advocates.

People's worldviews in this sensitive area hold an inordinate sway on their thinking. To take an extreme example, consider the many differences between men and women. We can't ignore the obvious physical differences — nor the more obscure ones, such as aspects of brain anatomy — and their genetic basis is indisputable. Clear behavioral differences are present too, although the relative influence of nature and nurture is uncertain here. Surely no one could imagine that men and women have *no* genetically based personality differences. Yet some people assert just that, saying that all such differences are cultural.

Unraveling the genetic differences among human populations sharing common ancestries may be difficult, but such differences also exist. Concern is so great about our ability to deal with the potential implications that investigators using data from the primary database on human genetic diversity at the NIH are required to sign a form stating they will not try to determine the ethnicity of the people who donated the samples. Although every one of the world's top two hundred times in the hundred-meter dash is held by someone of West African ancestry, it is commonly asserted that black athletic dominance of specific sports has nothing to do with genetics. So heightened are our political sensitivities about racial matters that some say the possibility of a genetic link should not even be examined.

Sir Roger Bannister, who broke the four-minute-mile mark in 1954 and is a retired Oxford dean, was sharply criticized when, at the 1995 meeting of the British Association for the Advancement of Science, he said, "As a scientist rather than a sociologist, I am prepared to risk political incorrectness by drawing attention to the seemingly obvious but understressed fact that black sprinters and black athletes in general all seem to have certain natural anatomical advantages." Theresa Marteau, of the Psychology and Genetics Research Unit at Guy's Hospital in London, responded, "It is potentially racist to look at the biological factors. I don't need to know whether what Bannister said is correct. And I don't think there needs to be research."

Avoidance is no solution. Genetics is clearly a key ingredient in athletic performance, as it is in other areas, and genetic differences among populations long isolated reproductively may cause a shift in the average potential of individuals in those populations. But generalizations about "blacks" and "whites" oversimplify and distort matters by suggesting some underlying unity within these categories. The broad groupings defined by superficial attributes like skin color include many distinct populations, each with its own common ancestry, and many people with mixed lineages. The genetic differences among ten black Africans from different parts of that continent, for instance, are far greater than the differences among ten light-skinned individuals from scattered places around the world.

Such sensitive issues will not remain in limbo much longer. Research will show that the influence of our genes on some human attributes is unclear or too hard to decipher, but other influences may be straightforward. The answers will be just another byproduct of the Human Genome Project. How we respond to this new information will be one of the biggest social and intellectual challenges of coming decades, for we will learn a great deal about ourselves that many people would rather not face. To date, most discussions of genetic information have focused on such issues as genetic privacy and whether to allow genetic testing for untreatable diseases like Huntington's. This is just the tip of the iceberg. Wait until the price of DNA diagnostic chips drops enough so that comprehensive testing becomes routine.

With the identification of every human gene, as well as the most common variants of each — the so-called single-nucleotide polymorphisms, or SNPs — we will be able to probe our genetics as never before. The key will be the DNA chip. In 1997, the company Affymetrix released the first such device, a dense gridlike array of 50,000 short DNA sequence probes on a glass chip. By mid-2000, when the rough draft of the human genome was completed, 400,000 such probes on a single chip were available for a few thousand dollars, and by early 2002, Affymetrix was selling researchers even better chips for as little as $200 each.

Dozens of large and small companies are working on chips designed to analyze tens of thousands of genes at a time. The result will almost certainly be tests that are fast, comprehensive, cheap, and reliable enough to read DNA sequences from a smear of saliva or blood. With this coming generation of gene chips, today's efforts to track changing levels of gene expression in tumors and other tissues will expand into broad population studies that bring new insights into our genetic makeup.

There will be technical and computational obstacles to overcome in characterizing our individual genomes cheaply and comprehensively, but emerging technologies do not need to analyze our individual genomes completely. Subtle effects can wait. Tests for the several million already identified common variants of our genes will yield enough information to keep researchers, epidemiologists, and clinicians busy while analytic techniques improve.

The burgeoning computational field of bio-informatics will be critical to this effort. No single human scientist will ever examine more than a tiny portion of our genome directly; it contains too many bases. Without computers to sift through this vast mass of data to find useful correlations between our genes and key aspects of who we are, the Human Genome Project would offer us a book we could never fathom. These computer tools are also essential to sophisticated embryo screening or germline manipulation because without the knowledge emerging from the Human Genome Project, few genetic interventions could be considered.

Those making sense of the output of the project don't care about germline manipulation — or embryo screening, for that matter — they are figuring out how to cure diseases and understand the workings of life. And there are fortunes to be made as well. Craig Venter not only made an enormous contribution to medical science by speeding up genome sequencing, he went from an NIH research salary to a net worth of more than $100 million in a few years.

Scientists are not going to shy away from their roles in this grand exploration out of fear that someone might engage in questionable human engineering someday. And we wouldn't want them to, because their progress will bring us enormous benefits, which is why the advent of human germline technology is virtually inevitable. Many interventions will prove too difficult, of course, but others may be surprisingly easy. Almost no one imagined in 1985 that the mutation of a single gene in a worm — the classic research organism *Caenorhabditis elegans* — would more than double its lifespan. For now, we can only guess whether we will unravel and control the biology of human aging or will find the knot too intricate to untie. But we likely will know soon.

Human Medicine

In his studies of Crigler-Najjar syndrome, Michael Blaese used a short stretch of specially tailored DNA to correct a one- or two-base mutation error in a gene. Introducing many copies of this molecule into a cell seemed to trigger its DNA repair mechanism to correct the targeted mutation. Blaese reported curing laboratory rats bred to have Crigler-Najjar, and if he succeeds in human studies, the breakthrough will be immense. This and other such work is not at the fringe of medical science, nor part of a plan to alter human reproduction, but cells are cells. If we learn to alter the genes of a liver cell more easily, we will probably be able to do the same in embryonic cells. Human politics does not trump the unity of biology.

The push toward therapies for adults cannot help but contribute to the development of germline technologies, but pharmaceutical development may prove even more important. While tens of millions of dollars each year go to gene-therapy research, billions will flow into pharmacogenetics — the effort to tailor pharmaceutical interventions to people's individual genetic constitutions.

Pharmaceutical companies estimate that they spend an average of seven years and $400 million to go from the discovery of a promising compound to an approved drug. The cost is so high because about 80 percent of this expenditure is on drug candidates that never reach market. The 1997 approval of the diabetes drug Rizulin shows what can go wrong. Although the drug initially seemed a big success, when a scattering of patients began to suffer liver damage and some died, calls to withdraw it mounted and Rizulin was doomed. The deaths do not mean that the drug didn't help anyone. Today, a drug is worthless if it benefits some patients and causes severe reactions in others. If these two groups could be distinguished in advance, however, the story would be different.

The hope of the pharmaceutical industry is that our genes will be the key to such predictions, and that as correlations are found between people's genetic constitutions and their medical histories and drug reactions, their drugs can be personalized. This and other medical hopes are helping fuel efforts to collect massive amounts of genomic data.

Kari Stefansson, the founder and CEO of deCODE Genetics, in Reykjavík, Iceland, has moved aggressively to uncover what the genes of Icelanders can tell us about disease. Until in-depth genetic surveys become affordable, deCODE must lean heavily on family histories, which makes Iceland an ideal place for prospecting. It has detailed medical records that go back to 1915, century-old genealogical records, and a homogeneous population descended from small bands of tenth-century Norse and Celtic settlers. In December 1998, Stefansson reached an unprecedented agreement with the

Icelandic government to link the country's health-care records with genealogical information about the island's 275,000 inhabitants, and license the information to deCODE. Three quarters of the population supported the initiative, and many Icelanders have given blood samples to provide detailed genetic information for the studies. The pharmaceutical company Hoffmann–La Roche was impressed enough to sign an agreement specifying research milestone payments totaling up to $200 million.

Such information helps find the genes implicated in various diseases, but if DNA chip technology moves quickly, large-scale genetic surveys that ignore genealogy will largely supersede Stefansson's approach. Researchers will depend less on finding disease genes in well-studied families and more on comparing genetic test results from tens of thousands of people who have a particular disease. Collaborative Genomics and other companies are already collecting hundreds of thousands of patient histories and tissue to prepare for such efforts. The government of Estonia is preparing to assemble genomics data for research on its 1.4 million citizens; the island nation of Tonga has signed an agreement with an Australian genomics company to compile data on its 100,000 inhabitants; and a company called Lifetree Technologies is collecting genetic samples from funeral homes and cremation societies.

Identifying a gene's association with a disease, however, is only the first step — a pointer to a link in a network of biochemical reactions connected with the observed condition. The hard task is teasing apart the workings of such pathways and figuring out how to manipulate them by pharmaceutical or genetic means. Such medical insights will be the initial harvest from genomics, but as we unravel our underlying biology, we will draw ever closer to finding genetic modifications to manipulate it. Ultimately, this knowledge will be the foundation of efforts to expand human biological capacities through germline interventions.

Basic and Applied Animal Research

Wide-ranging research in academic and corporate laboratories, stoked by rapid technological advances in instrumentation and computer technology, is driving progress in molecular genetics. The motivations of basic researchers elucidating human biology range from intellectual curiosity and visions of helping humanity to scientific competitiveness, winning job promotions, and forming start-up companies. Redesigning our children's genetics isn't on the list.

In early 2001, I spoke with Gerald Schatten, the soft-spoken investigator at the Oregon Regional Primate Research Center who created the first transgenic primate, ANDi, a rhesus monkey. Schatten is working on cures for human disease. He strongly opposes human germline manipulations, and his primate procedures are unsuitable for use in humans, but the arrival of ANDi was yet another step toward eventual human germline interventions.

Similarly, when Mario Capecchi, a distinguished geneticist at the University of Utah, the recipient of the highly regarded Kyoto Prize, and a former student of James Watson, set out to develop methods for making targeted genetic changes in the germinal cells of mice, he did not have human manipulations in mind. He was trying to devise a way to modify mouse genes selectively so he could study them better. His pioneering technique for creating "knock-out" and "knock-in" mice by inactivating or inserting genes is now a cornerstone of genetic research and is so widely used that a researcher can contract for a new knock-out strain for about $7,000.

Capecchi's "knock-out" methodology is far from trivial. A researcher wishing to induce a specific germline modification in a mouse gene injects the mouse's fertilized eggs with sequenced DNA encoding the intended change. Once every few hundred eggs, the injected DNA replaces the original DNA in a nucleus and creates the desired genome. After verifying that this has occurred, researchers grow modified embryonic cells and implant them in normal mouse embryos. The mouse pups born from these hybrid embryos have patches of differing cells, some with unaltered genes, others with the knock-out gene. By breeding mice whose sex cells belong to a patch of cells containing the knock-out gene, researchers can produce a pure knock-out strain in the next generation. These experiments take about a year to complete.

Such breeding is obviously unsuitable for germline modifications to people, but researchers nonetheless can study a human gene by knocking it into mice or by knocking out the equivalent mouse gene. This is how medical researchers create strains of mice, called mouse models, that mimic human disease. For example, by knocking out the gene for a protein that influences nuclear gene expression, researchers at the University of Texas Southwestern Medical Center created a mouse strain that consistently develops colorectal cancer resembling that in humans. Such models are extremely valuable, and researchers have constructed them for Alzheimer's, inflammatory bowel disease, cystic fibrosis, juvenile diabetes, and other diseases.

As this research elucidates the workings of our genes, our bodies, and our minds, the day draws nearer when we will be able to select and manipulate the genetics of our children. Human germline procedures will piggyback on such research, in the same way that the possibility of human cloning emerged inadvertently from work on pharmaceuticals.

Besides the little lamb that Mary had — the one whose fleece was white as snow — Dolly, with her endearing little Finn Dorset face, is probably the most famous lamb of all time. Born at the Roslin Institute in Scotland on July 5, 1996, she was the first mammal cloned from an adult cell. Her birth announcement the next February rocked the world by making people confront the possibility that humans too would one day be cloned. Many critics dismissed the possibility, emphasizing the difficulty of the procedure and pointing out that scientists needed 277 embryos to obtain a single lamb. But the public was unconvinced, and with good reason. The genie was out of the bottle.

Successful clonings of mice, cows, goats, and pigs followed, well-funded attempts to clone cats and dogs began, and less than four years after the announcement of Dolly's birth, lo and behold, a fertility specialist, Dr. Panos Zavos, was testifying before the U.S. Congress that he would clone a human.

But Dolly's creators at Roslin and their partners at PPL Therapeutics were not working toward this end. Cloning was but a step in refining transgenic technology in order to put human genes into sheep and quickly build viable transgenic flocks that would secrete human proteins into their milk. In a single generation, one sheep could give rise to a large enough flock to produce pharmaceutical-laden milk in quantity.

Using dairy animals to produce pharmaceuticals instead of plain old milk is known as pharming, and PPL along with two rivals have been considering dozens of proteins for large-scale production. PPL's U.S. subsidiary in Virginia has created cows whose milk brims with a human protein that is a nutritional supplement for babies.

Until recently, companies extracted clotting factors for hemophiliacs from blood plasma. But this was expensive, and given that more than half the hemophiliacs treated between 1977 and 1985 contracted AIDS from tainted plasma, it also posed a danger of transmitting blood-borne viruses. Extracting the factors from the milk of transgenic animals could be cheaper and safer, and at the end of 1997, scientists used cloning technology to create Polly, a ewe with the gene for factor IX, which is used to treat hemophilia B.

The world market for this clotting factor is around $160 million a year. That for human serum albumin, which is used to treat burns, may be ten times higher. Commercial possibilities such as these — not applications in human reproduction — are driving the development of cloning and transgenic technologies. Not only have Ian Wilmut and other key figures at Roslin and PPL spoken forcefully about the immorality of human cloning, no lure of cloning profits exists to tempt them to do otherwise, because the technology is far from being safe for humans, much less commercially viable. That Roslin had mentioned human applications in its patents does not indicate some sinister plot to produce human clones. Such handling of intellectual property is standard.

The companies doing transgenic work in livestock are among the harshest critics of eventual human germline procedures, but however adamantly and genuinely they oppose these applications, they are laying their technical foundation. When I sought funding in 1997 for the symposium Engineering the Human Germline, I approached the Biotech Industry Organization, naively thinking they might support the forum, the first large public discussion of this controversial topic. Not only did they say no, they urged me not to hold the event and told me that what I was attempting was highly irresponsible because it might create a backlash against important biotechnology.

Human Infertility

Another field helping to build a foundation for human germline intervention is that devoted to the treatment of infertility. No one here cares about such wild notions as human redesign; everyone is too busy counseling patients, performing ultrasounds, aspirating eggs, overseeing lab procedures, and implanting embryos. They are occupied with the here and now — with women and men who cannot have the babies they badly want. The larger import of their work, however, is unmistakable.

When Louise Brown was born by *in vitro* fertilization in 1978, many observers greeted her with considerable handwringing as a "test-tube baby." Jeremy Rifkin, a perennial opponent of new technology, went so far as to suggest that such a child might be psychologically "monstrous." "What are the psychological implications of growing up as a specimen," he wrote, "sheltered not by a warm womb but by steel and glass, belonging to no one but the lab technician who joined together sperm and egg? In a world already populated with people with identity crises, what's the personal identity of a test-tube baby?" Nearly a generation later, we have no trouble answering his question: a child of IVF is the same as any other child.

In vitro fertilization is now the choice of tens of thousands of couples who would not otherwise be able to have children, and in some clinics, the success rate for women under thirty-five is more than 70 percent per ovulatory cycle, much higher than for natural conception. In the United States in 1998, 28,000 babies were born this way, and doctors performed 80,000 cycles. The biggest complaint is not that IVF is immoral, but that it costs so much and often doesn't work. An IVF cycle in the United States typically costs about $10,000 and is not covered by health insurance.

No matter what methods eventually arise for shaping the genetics of human embryos, IVF procedures — the extraction of eggs, their *in vitro* fertilization, and the implantation of the resultant embryos — will underlie these methods. So present efforts to refine and enhance IVF procedures are also working toward germline interventions.

IVF works well today, but it remains too expensive, unpleasant, unreliable, and intrusive to compete head on with good old-fashioned sex. IVF is a tool to treat infertility, and even for that, couples often choose it as a last resort.

Headline stories about Britain's oldest mother, sixty-year-old Elisabeth Buttle, or Italy's sixty-nine-year-old Rosanna della Corte may lull many women into delaying childbearing until they approach their forties, but by then IVF is quite problematic. *People* magazine's happy story, in June 2000, about the experience of Cheryl Tiegs, then fifty-two years old, with a surrogate mother is even more misleading about the difficulties facing older women seeking IVF:

> Tiegs underwent fertility treatment to increase the number of eggs that could be harvested and then fertilized. The resulting embryo was transferred to the surrogate's uterus, and the procedure was successful on the very first try. "It's my egg and my husband's sperm, so they're our babies," says Tiegs . . . The couple are eagerly awaiting the day in July when they'll rush to the hospital to be with the surrogate when she gives birth to their fraternal-twin boys.

A mistaken story of this sort would be harmless enough, except for its influence on those thirty-somethings mulling whether to have a baby now or wait a little longer. Given the medical hype about fertility procedures, many women might imagine they can easily have a child using their own eggs well into their forties. It just isn't so. The forty-four-year-old woman I know who innocently mentioned to me one day that she was starting to think about having a child by IVF (using her own eggs, of course) was sadly deluded. Even by thirty-five, a woman's chances of successful IVF with her own eggs are beginning to fall. By forty her chances are 15 percent per cycle, by forty-two 8 percent, and by forty-four so small —

less than 1 in 30 — that many clinics will not even try. By fifty we're entering the realm of a medical miracle. For a woman of fifty-two to conceive a viable pregnancy using her own eggs would be astonishing; for her to have fraternal twins on her first try is hardly believable.

To make the underlying realities of IVF more concrete, here is a typical cycle that a close friend of mine went through. Following a battery of unpleasant tests a week before her period, she started daily injections of a drug to prevent premature release of eggs. On the second day of her cycle, she began injections of a hormone as well, to hyperstimulate her ovaries to produce extra egg follicles. Ten days later, when ultrasound showed that a number of the follicles had grown to a diameter of about three quarters of an inch, she stopped both injections and triggered egg maturation with a shot of another hormone. Exactly thirty-six hours later, her doctor put her under general anesthesia and extracted the fluid from each follicle by passing a needle through her vaginal wall and into her swollen ovaries. A technician put the fluid in a petri dish, found nine tiny ripened eggs, and fertilized them with fresh sperm that her husband had just deposited in a plastic beaker while watching pornographic videos in a bathroom down the hall. (As in normal reproduction, men have it easy.) Three days later, the doctor used a thin plastic tube to place the three healthiest-looking embryos, each containing about eight cells, into her uterus, and she began three weeks of still more shots, this time to thicken her uterine lining so the embryos would have a better chance of successfully implanting. Only at the end of this period did a pregnancy test reveal that it had all been for naught. None had implanted.

Given this ordeal and the mood swings brought on by the hormones, no wonder IVF still accounts for fewer than 1 percent of live births in the United States. Improvements, however, may transform the procedure enough to integrate it into routine procreation. With a little marketing by IVF clinics, traditional reproduction may begin to seem antiquated, if not downright irresponsible. One day, people may view sex as essentially recreational, and conception as something best done in the laboratory.

Three technologies will likely combine to convert IVF into a commonplace reproductive procedure that is suitable for germline interventions. The first will be the maturation of immature eggs retrieved by simple ovarian biopsy. This will raise the number of eggs a woman might use for reproduction and eliminate the need for heavy doses of hormones to stimulate her ovaries. The second will be the freezing and thawing of immature eggs without damaging them. This will allow a young woman to bank healthy eggs without having to decide then and there whose sperm will eventually fertilize them. She could thaw eggs from this external artificial ovary anytime, then mature and fertilize them to make large numbers of embryos. The third technology is comprehensive genetic testing of embryos. Such nondamaging testing would allow couples to make meaningful choices about the embryos they implant.

Genetic testing of embryos is nothing new. Physicians first performed preimplantation genetic diagnosis (PGD) in 1989 at London's Hammersmith Hospital by teasing a single cell from each of several eight-cell embryos and testing the gender of the cells, so they could implant a female embryo and avoid a sex-linked disorder that occurred only in males. Two years later, the same physicians tested for a genetic mutation that causes cystic fibrosis and enabled a high-risk couple to implant an embryo free of the disease. Today, couples can use PGD to screen for a handful of genetic diseases, including hemophilia, Tay-Sachs, and fragile X syndrome, which can cause severe mental retardation in males. Because of high cost, limited availability, and the small number of conditions that can be tested, the procedure is still rare, but in a decade the landscape will differ. Scientists will have identified associations between constellations of genes and various physical and mental attributes, including the risks for common diseases. Couple this with the future use of PGD to look at many more genes in a single embryo, and we will probably be able to test an embryo and obtain solid information about the child it would become. Parents will then have a choice as to which potential child they'd like to bring into being.

The usefulness of such embryo testing will depend on the extent to which our genes shape us and on the complexity of their influences. Some important aspects of who we are will turn out to be largely independent of our genes; others will have strong genetic contributions. At present, our knowledge of this is only general. Lung cancer, for example, seems to depend mostly on environmental influences, while prostate cancer appears to be more influenced by genetic predisposition. In a decade or so, however, we should have solid, specific answers about many of these relationships.

We are the result of an intricate interplay of genes and environment, and the two are interdependent. Our genetic tendencies can shape our environment by steering our choices, and environmental influences can switch genes on or off. This means that as society ever more successfully eliminates extreme variations in environment — say, by providing basic nutrition and education to all — genes will become more, not less, important influences in shaping us. As we correlate our genetic constitutions with the details of our health, personality, and behavior, we will learn not what our genes determine in some absolute sense, but what possibilities they push us toward within the normal range of environments we encounter. Deciphering the workings of our biology and the influences of experience and environment will give us more control over our lives by offering us more effective ways of diminishing the vulnerabilities and adding to the potentials that our genes bring us.

The Market: Violator or Protector?

That no one today is attempting or even seriously contemplating engineering the genes of human embryos is not surprising. After a decade of well-funded work, physicians still cannot give corrective genes reliably to patients suffering from the single-gene disorders we understand, much less the multiple-gene ones we don't. But the capabilities of sophisticated embryo selection and human germline technology will flow directly from discoveries in the mainstream research discussed here. If someone somewhere were to design a crash program to bring us safe and reliable germline interventions, its outline would be remarkably similar to what is going on today.

As for the hesitance of medical practitioners to embrace the concept of intentional human redesign, anything else would be astonishing. Such work is oriented toward future generations, not those who are alive now and crying out for care.

Germline selection and manipulation lie beyond medicine's present boundaries, but these boundaries may shift rapidly in response to public interest as real opportunities emerge. In a 1993 international poll, Daryl Macer, the director of the Eubios Ethics Institute in Japan, found that a substantial segment of the population of every country polled said they would use genetic engineering both to prevent disease and to improve the physical and mental capacities inherited by their children. The numbers ranged from 22 percent in Israel and 43 percent in the United States to 63 percent in India and 83 percent in Thailand.

Speaking to a pollster about abstract choices is one thing; making concrete decisions about our children's genes is another. But widespread use of coming reproductive technologies is not hard to imagine. Large numbers of young women would likely bank eggs if they could do so easily. If nothing else, that would calm the angst about their biological clocks running out. Many such women, of course, would never use their banked eggs; they would conceive their children through sex. But other women would choose embryo implantation, seeing it as a trivial procedure too good to pass up. Some couples would find the option of timing the conception and birth of their child compelling. Others would want to screen for worrisome genetic diseases or eliminate the spontaneous abortions caused by genetic abnormalities. Still others would want to select some attribute of their future baby: gender, adult height, hair color, or temperament. The motivations behind such choices would be as personal and varied as the lifestyles and values of the couples.

Women who have banked eggs would not be the only ones to avail themselves of these options. Millions of couples have infertility problems. The numbers of people who use IVF would swell when the procedure became less expensive and onerous, and they too would find such choices enticing. Also, as IVF became easier, many women who never bothered to freeze eggs might decide to use the technology simply for the choices it offered them.

The procedure would be straightforward. Typically, a woman would enter an IVF clinic as she does today, but instead of facing an exhaustive ordeal, she would have an ovarian biopsy to obtain fresh eggs or would just send for her previously frozen eggs. Her partner would deposit sperm as he does today or send for previously frozen sperm. In light of the recently identified risks in conceptions by men over forty, such sperm banking could well become increasingly common. Conception would occur in the laboratory, and the woman would later return to have the growing embryo implanted.

For "natural" conception, the woman would implant one of the embryos at random. For disease screening, she would implant one that had passed a genetic test for chromosomal abnormalities and disease mutations. For embryo selection, she would implant the one whose PGD results best matched the genetic predispositions she and her partner had chosen. For germline intervention, she would also implant a selected embryo, but one whose genes had been modified.

As for regulating the use of advanced reproductive technologies, a tension exists between those who advocate tight government control and those who believe that with a modest amount of oversight, the free market is best suited to determine their use. Those who oppose government control often argue that denying a personal reproductive right is inappropriate. This "right," however, is more a simple assertion than anything else. These new reproductive possibilities are not enshrined in law.

But setting aside this issue, consider the extent to which the free market protects us from new technologies we do not want. The market is a very powerful influence in human affairs, and whether or not these technologies are legally sanctioned, commerce will almost certainly help shape their future. A major safeguard that capitalism offers against questionable reproductive technologies is that unless they appeal to a significant number of people, the procedures have no profit potential and fade away or are not even developed. This also means that of those reproductive and genetic technologies that are reasonably safe and reliable, the ones we see first will likely be those with large potential markets.

Somatic gene therapy gives us a concrete example of how market forces determine the targets of medical research. Tay-Sachs and cystic fibrosis both result from single defective genes. Both are horrible. Both are fatal. But while a huge amount of work has gone into developing a treatment for cystic fibrosis, the Tay-Sachs effort has been modest. A big factor is that in the general population, Tay-Sachs disease is rare, striking fewer than two dozen babies a year and killing them by the age of five, while forty thousand people suffer from cystic fibrosis in the United States alone, and the average annual cost of their health care is $25,000. Whether private pockets or public coffers pay for gene therapy for cystic fibrosis, the treatment could tap into the billion dollars or so presently being spent for care. A therapy for Tay-Sachs has no such fund to look to, however, and would not recoup its development costs, so progress has to piggyback on other work.

Abuses of germline interventions and other advanced reproductive technologies in an unregulated environment are not nearly as likely as some critics imagine, because strange fringe interventions would have too few users to command the funds required to bring them into open use. Human germline manipulation, for example, could never hope to become technically feasible except as a spinoff of huge research projects in related arenas. Any financial payoff from it is too distant, liability issues are too worrisome, and development costs are too high. Even as a spinoff, germline manipulation would probably not reach the open market until an extremely desirable enhancement such as extended lifespan was demonstrated.

An even bigger protection is the freedom to forgo any technology the free market offers. As long as the government does not force people to use these reproductive technologies, people are at almost no personal risk from them. Some, of course, maintain that the public doesn't know what is best for itself, and that free parental choice is too dangerous. While our choices are not always wise, the assertion invites an obvious question: who is in a better position than parents to make such determinations about so intimate an aspect of our lives as having children?

When people have a range of reproductive options, they generally try to get what they want in the easiest, cheapest, safest, and most reversible way. No couple would use germline interventions to protect their future child from polio, knowing they could later vaccinate the youngster instead. Nor would a couple worry much about a child's potential risk factors for diabetes if they knew the disease was, or soon would be, curable. Germline therapy for genetic diseases like Huntington's or cystic fibrosis is also unlikely, because embryo selection using PGD will be a much better way of avoiding these conditions. Parents philosophically opposed to diagnostic screening might wish to repair a genetically flawed embryo they created during IVF, but I suspect that more such people exist in the minds of clever bioethicists than in the real world. Germline therapy would bring the destruction of embryos as surely as embryo screening does.

Human germline manipulation will come into being not as a replacement for existing technologies like embryo screening, but as an extension of them. Germline modifications will appeal to us to the extent that they can deliver compelling benefits that we cannot obtain using simpler procedures. Direct germline intervention is the logical conclusion of our ongoing progress in reproductive biology and the ultimate expression of it, and its realm will likely be human enhancement.

4

Superbiology

> "Cheshire Puss," she began . . . "Would you tell me, please, which way I ought to go from here?"
>
> "That depends a good deal on where you want to get to," said the Cat.
>
> "I don't much care where —" said Alice.
>
> "Then it doesn't matter which way you go," said the Cat.
>
> — Lewis Carroll, *Alice in Wonderland*

SPEAKING LOOSELY about "designer children" and how people might one day give kids genes to enhance their intelligence, beauty, or athletic ability is one matter; actually making these interventions is quite another. Anyone in medicine or the life sciences knows the complexity of biological systems and the challenge in establishing what observable characteristics some constellation of genes will produce, especially when set against different genetic backgrounds. For example, at age seventy, about half the women who have a mutated BRCA gene but no close family members with breast cancer have not contracted the disease. Some combination of genes, environment, and luck has protected them, but scientists cannot yet say which factors are most important for a specific individual.

Without a doubt, bizarre notions of giving people gills or wings are the stuff of science fiction. Figuring out the impact of a single gene designed *de novo* or one taken from another species would be daunting enough, yet the task is trivial compared with designing entirely new capacities orchestrated by many genes. Moreover, human interventions must be without questionable side effects. With laboratory mice, failures are part of the process, and procedures can be refined over the years, but human design must have a good chance of success from the outset. These challenges, together with the possibilities of embryo selection, have led some to dismiss human germline manipulation, even for seductive enhancements to extend lifespan or improve health. This notion is mistaken. It gauges the difficulty of genetic interventions by the rudimentary gene transfer technologies of today, and it ignores the rapid growth of knowledge about our genes.

Our coming capacity to manipulate our genes will not be defined by our ignorance of the many genes and gene combinations we do not understand, but by our depth of knowledge about the few we do understand. These will be our first targets. As we unravel our genetics, we will find that many traits are too opaque for us to manipulate anytime soon, that others are somewhat obscure but seem feasible in the near future, and that still others are straightforward. We do not yet know which human traits and conditions lie in which category, but within a decade or so we should have a fair idea of the size of the task facing future genetic engineers.

The apparent nature of a trait offers little insight into the complexity of the genetics underlying it. Perfect pitch is the ability to identify a musical note without any external reference for comparison. The physiological and cognitive mechanisms for accomplishing this are no doubt complicated, so one might expect that the ability would have numerous genetic contributions, but family association studies suggest that the potential to develop perfect pitch may hinge on only a single allele (genetic variant of a particular gene). The realization of the skill, however, correlates with early musical training, typically beginning before age four, so perfect pitch, like so many other abilities, requires both a genetic predisposition and early training.

Before we look at the diverse physical and mental capacities that might be tempting targets for genetic manipulation, we need to consider how future genetic engineers might go about the task.

Prospective parents will not seriously consider altering the genetics of their future children until there are both seductive genetic constructs to use and safe, reliable methods to put them into an embryo. Neither exists today. No present genetic intervention is worth doing in a healthy individual, and no present technology is capable of effecting an intervention safely anyway.

The techniques now used to manipulate animal germlines tell us little about how future germline interventions in human beings will be accomplished, because the animal methods are so ill suited to human use. When Polly, the sheep with the gene for human clotting factor IX, was born in 1997, the event vindicated the technologies pioneered with Dolly, but growing numbers of such transgenic manipulations of livestock will not segue into human modifications. The incentive for a physician to attempt an equivalent human intervention — perhaps just for the sheer notoriety of being first — exists, of course. But even if some present-day Dr. Moreau were to successfully alter the genetics of a human embryo this way, a wide gulf would still stretch between clinical procedures suitable for general use and the large, expensive effort required for this first success.

Using livestock methods in humans would be risky, cost a small fortune, and open doctors to the possibility of damaging lawsuits. Couple these problems with the difficulty of recruiting surrogate mothers and medical staff for a procedure that buys an occasional success with hundreds of failures, and the obvious conclusion is that even in a completely permissive legal environment, these methods would have to improve dramatically before they could become real options for people. As for the breeding methods used to create genetically modified mice, the tight controls over mating they require make them even less suited for human use.

The inadequacy of current research protocols for human use has colored today's discussions of human germline manipulation, reducing them to vague futuristic predictions, in *Time* magazine and elsewhere, of "designer babies." Not surprisingly, attempts to come to grips with the ethical, social, and scientific dilemmas arising from fanciful interventions performed with inadequate technologies have not led very far. But technological barriers soon may fall.

In order to better comprehend the issue, let's explore one approach geneticists might employ to reliably place genetic elements in embryos.

Although natural conception itself is risky, a germline technology may have to be close to foolproof to figure prominently in humanity's future. As indicated earlier, we are highly intolerant of risk in processes we control. Germline technology would also have to be cheap and simple enough for widespread use. If meaningful genetic manipulation, for example, was possible technically but depended on a series of laborious interventions that had to be tailored to each individual genome, a mini–research project would have to be designed to plan each embryo's modifications. Until such procedures became effective at creating truly exceptional humans, they would not be very influential.

Auxiliary Chromosomes

Widespread modifications of human genetics will require reliable generalized methods for germline intervention. The human artificial chromosome, pioneered by John Harrington and Huntington Willard in 1997 at Case Western Reserve University, has that potential. This technological descendant of the bacterial artificial chromosome and the yeast artificial chromosome, which have been in use for many years, contains all the essential elements for chromosome functioning. In 2001, two companies were at the vanguard in developing this technology.

Back in 1998, one reported that its synthetic chromosomes had passed stably through more than a hundred cell generations in human tissue culture. And in 1999, the other announced that its artificial chromosomes had been retained by successive generations of mice reproducing normally. The implications of the mouse result for human germline interventions were so clear that the inventors felt compelled to state that they planned to use the technology only for animal transgenic work and human somatic therapy, and would not license the chromosomes for human germline therapy. It is doubtful that this will have any more long-term impact on the use of the technology than statements from the Roslin Institute opposing human cloning.

Adding a new chromosome pair (numbers 47 and 48) to our genome would open up new possibilities for human genetic manipulation. The advantages of putting a new genetic module on a well-characterized artificial chromosome instead of trying to modify the genes on one of our present 46 chromosomes are immense. Not only could geneticists add much larger amounts of genetic material, which would mean far better gene regulation, they could more easily test to ensure that the genes were placed properly and functioning correctly.

Because an artificial chromosome provides a reproducible platform for adding genetic material to cells, it promises to transform gene therapy from the hit-and-miss methods of today into the predictable, reliable procedure that human germline manipulation will demand.

Ideally, an unloaded auxiliary human chromosome would have no functional genes of its own. It would be an inert scaffolding dotted with independent insertion sites where modules of genes and their control sequences could be placed using the various enzymes that splice and clip DNA. With adequately separated sites, the eventual payloads of genetic modules would not interact with one another and could be expressed independently. The auxiliary chromosome would be a universal delivery vehicle for gene modules fashioned by medical geneticists throughout the world. At first, only a few safe and effective modules would exist, primarily those specific constellations of gene variants known to confer clear advantages because they occur naturally and have obvious benefits. Eventually, geneticists might develop hundreds of such modules, each with its own particular benefits and risks.

Delivering gene modules using synthetic chromosomes would be the safest and least intrusive way to substantially modify our genetics. By not altering a single one of the 3 billion bases on our existing chromosomes, geneticists would minimize the chance of inadvertently stepping on the many yet unappreciated interactions within our genome. Given the limits of our knowledge, however, we would be wise to design any insertion sites to include a mechanism for selectively switching off the expression (that is, the activity) of the genetic module placed there. An injection could provide the chemical signal that would trigger the shut-off.

Auxiliary chromosomes may not be the only way of achieving workable, complex germline manipulation, but they give us an idea of the level of sophistication we may one day attain in modifying human genetics. At first glance, the framework I've sketched of precisely regulated human artificial chromosomes may seem reminiscent of the grand schemes of the artificial-intelligence experts whose cyborg visions appear to be out of touch with reality. But rudimentary artificial chromosomes exist. Efforts are well under way to insert and express genes in them. The enzymes to edit DNA have been copied from living cells and are in widespread laboratory use. And our genomes already possess finely tuned genetic controls that turn our genes on and off at the right time and place. These control sequences, crafted by the unforgiving process of evolution, work exceptionally well, and they will give us more than just templates that future genetic engineers can study to develop new designs; when copied — as already occurs in mouse research — these controls will offer many of the precise patterns of gene expression that are needed.

The earliest genetic modules for artificial chromosomes will probably imitate nature closely enough not to require inordinately more knowledge than we are currently gaining. Chip-based gene surveys of large numbers of people will identify constellations of gene variants that seem valuable. Geneticists will copy these variants and couple them with existing regulatory sequences. Some of these genes may work well even though they represent an extra copy beyond the maternal and paternal versions we normally possess; others may not. For example, a child with an extra copy of our tiniest chromosome, number 21, with a mere 225 genes, has Down syndrome, and any other extra chromosome except X or Y is usually fatal.

To avoid the imbalances that replacement genes might cause will require care. If geneticists replace a gene by adding a new version of it to an artificial chromosome, they will likely have to "silence" the original gene. Researchers accomplish this today by interfering with steps in the process by which a particular gene is translated into the protein it encodes.

As long as a gene is regulated properly, relocating it to a new chromosome would probably not be a problem. Chromosome rearrangements occur frequently in nature. Our own chromosome 2, for instance, arose from the fusion of two ancestral chimpanzee chromosomes some 5 million years ago. A comparison of the banding patterns on the relevant chromosomes shows clearly that this fusion is why we have 23 chromosome pairs instead of the chimp's 24.

Auxiliary chromosomes for human use are bound to gradually pick up other design features, bells and whistles that offer new potentials and protections. For example, once geneticists can build modules that do more than mimic gene combinations already found in the human population, parents may want to keep the new modules inactive until the recipient is old enough to decide whether to use them. Such gatekeeping makes no sense for genes expressed in fetal development or childhood, but genetic constructs that retard the onset of aging or protect us from adult cancers might come into play later in life. Such preinstalled midlife interventions might offer an individual the future option of activating otherwise impracticable gene therapies that precisely target various ills. At present, we understand early human development too poorly to confidently modify it anyway, but designing and controlling late-onset interventions in an adult would be much easier, and making their activation reversible would offer additional protection against unanticipated complications.

A technical design that allows a recipient to decide when to activate his or her genetic modules requires an element to keep the gene silent in the absence of some preselected chemical signal. Such a "lock" on gene expression sounds like science fiction, but it would be only a modest extension of the cell's existing ability to shut down whole suites of genes when it follows one developmental path instead of another.

The Purkinje cells, for example, are the largest and most complex of our nerve cells. Their cell bodies are nearly the diameter of a human hair, and their dense, tangled bushes of dendrites, the long fibers that connect to other cells, are the sole pathways out of the cerebral cortex and link up with as many as 200,000 other nerve cells. Many of the genes active in the Purkinje cells are inactive in other cells, such as those that synthesize collagen and other constituents of our bones. But both cell types have the same underlying genes.

Our cells can shut off selected suites of genes and also turn them on again. Many of the genes in the mammary-cell nucleus used to create Dolly, for example, were silent when they were in the mammary glands. The key to cloning was to develop ways of reversing this, so that the genes appropriate for embryonic development could become active when the nucleus was injected into the egg that later became Dolly.

Ways of regulating genes on an artificial chromosome so that the recipient can decide whether or not to turn them on are here for the taking. A system that uses tetracycline as the signal to turn specific genes on or off, for instance, has been in wide use in laboratory research for years. Artificial chromosomes will need other features too. Ideally, these constructs will be easy to handle, easy to load with gene modules, easy to monitor to verify their integrity, and easy to keep from passing to future generations.

Critics of human germline manipulation frequently point to the risks of passing on genetic errors to future generations. But even if errors are entirely preventable, making our early genetic modifications permanent parts of the human gene pool would be foolish. When children who have received auxiliary chromosomes to improve some mental or physical characteristic grow up, they may want to give their own child the same advantage. They won't, however, want to pass on the outdated auxiliary chromosomes they received a generation earlier, any more than a middle-aged father today would try to give his Internet-savvy college-bound daughter that state-of-the-art typewriter he used for his term papers. Parents will want the most up-to-date genetic modifications available. Were these prospective parents' own modifications scattered through their chromosomes, cleaning them out and upgrading them would be tricky, but with changes confined to an auxiliary chromosome, a parent could simply discard the entire thing and give his or her child a newer version.

One Generation at a Time

Casual comments about jettisoning a chromosome may sound glib, but Mario Capecchi, the scientist who pioneered the creation of knock-out mice in 1989, has already come up with a way of doing this. He first described it at the 1998 symposium Engineering the Human Germline. John Campbell, a professor of neurobiology at UCLA's School of Medicine, who organized the event with me, had been toying with various unsatisfactory schemes for keeping a chromosome from passing from parent to child, and we asked Capecchi if he could come up with a better method. He presented an elegant solution in a wonderful talk titled "Human Germline Gene Therapy: How and Why."

Capecchi's solution to heritability illustrates how misleading it can be to separate ethical discussions of a future technology like germline engineering from the technical details of its implementation. Prior to 1998, genetic manipulations in humans were assumed to be either somatic, and not heritable, or germline, and passed to future generations. Bioethicists almost uniformly declared that human germline alterations were wrong, because people did not have the right to shape untold future generations. A position piece written in the fall of 1992 by the Council for Responsible Genetics, for example, unconditionally opposed human germline engineering, calling for a ban on the practice. They claimed that the technology was "unnecessary to save the lives or alleviate the suffering of existing people," and that there would not be "accountability to individuals of future generations who are harmed or stigmatized by wrongful or unsuccessful germline modifications of their ancestors."

No bioethicists stopped to consider that single-generation germline manipulations might be possible, but Capecchi has now shown that we might modify the genetics of the first cell in an embryo to change every cell in a future being, yet program those changes to drop away at the time of reproduction and not flow to future generations. What could be more respectful of distant descendants than a genetic modification configured to allow them either to reaffirm their parents' vote for genetic redesign or to cast it off and reclaim their "natural" genome?

Capecchi's method relies on an enzyme called CRE, which snips out the DNA between two copies of a specific sequence called a *loxP* site and reconnects the loose ends. Placing one site on either side of a group of genes marks it for deletion when the CRE gene is turned on.

In this way, an entire auxiliary chromosome could be marked for later removal by bracketing its centromere, a region needed for proper chromosome handling during cell division. Geneticists would put the CRE gene, which does not exist naturally in humans, under controls that activate it only in sex cells and only in the presence of a preselected drug. In other words, we could take a pill to discard our auxiliary chromosome exclusively from our sex cells and thereby keep it and all the genes on it from reaching the next generation.

This is not pie-in-the-sky genetic design. Capecchi's lab has already used the technique to remove a test gene inserted in a mouse chromosome. The gene remained active in the mouse's body cells, disappeared from the sex cells as predicted, and failed to show up in tests on more than a hundred offspring. Nor is this targeted-deletion system unique; an equivalent but different one is in use for research on fruit flies. Combining such systems would allow multiple, independently controlled deletions in humans.

The various ideas I've described for human germline manipulation are by no means comprehensive, but they indicate the elegant approaches that artificial chromosomes offer. Moreover, a chromosomal platform of this sort would work in any animal, so clinical methods could be refined in tissue culture and animal embryos before human use. Ironically, such thorough testing may not be possible with somatic therapies, because they must rely on diverse methods that may not translate easily to experimental animals. A universal delivery vehicle for genetic interventions has obvious advantages for safety and reliability.

Combating Disease

John Campbell revels in theoretical evolutionary biology and has an incisive mind unfettered by one shred of political correctness. Indeed, he delights in stirring up a fuss, as he did with the iconoclastic lecture "Humans K–12," in which he suggested hiring a cadre of professional testers to try out pharmaceuticals that have been shown to be safe in animals and recording their reactions in elaborate detail. John considered the possibilities of human germline engineering years before the idea was being seriously discussed, and he has worked out various interventions that may become practical once geneticists are able to insert genes and their control elements on chromosomes. His design for a protective against AIDS illustrates how germline procedures using artificial chromosomes could build on existing research and target specific groups of cells.

The human immunodeficiency virus, which is responsible for AIDS, infects only certain cell types made by the bone marrow, most notably the T helper cells, or T cells, in our immune system. Since these cells are responsible for key aspects of immune response, the disease leaves people at the mercy of opportunistic infections they would otherwise fight off. AIDS researchers looking at how an AIDS infection occurs, how it spreads, and why rare individuals can resist the disease have made progress toward treatments and are examining a number of modified genes that might make engineered T cells resistant to the virus. Some possible therapies rely on inserting one or another of these genes into the bone-marrow stem cells of AIDS patients to produce resistant T cells, but the formation of mature T cells is complicated; engineering the bone marrow of an adult to produce resistant T cells is an ambitious goal. Introducing resistance genes into the germline to prevent AIDS might be easier.

If resistance genes are found, the most cautious way of introducing them into the germline would be to ensure that they were active only where needed — in this case, in the T cells. The way this could be accomplished illustrates how future germline interventions might target specific types of cells.

To understand how such targeting would work, we first need to look a little more closely at how genes are regulated. The promoter, a stretch of DNA immediately upstream from a gene's coding region, is one of the best understood of a gene's regulatory elements. Promoters serve as attachment sites for special protein factors that enable a promoter's associated gene to be expressed. Our genome codes for thousands of such factors, and the particular ones a cell makes determine which genes and groups of genes are active in that cell.

The virus that causes AIDS enters only those cells that make a protein called CD4, which resides on their surface, so CD4 is a marker for the cells that are vulnerable to AIDS. One strategy for regulating a gene for AIDS resistance is to paste a copy of the promoter for the marker gene in front of the resistance gene. That way, the resistance gene will be active only where the marker gene is, which is only in the T cells that need defending.

No matter that the marker's promoter is so long and intricate that it is not yet understood, despite a great deal of research attention. Scientists don't need to understand the promoter's workings in detail, because their goal is to mimic it, not decipher it. This targeting strategy is not unique to defending against AIDS; the approach is general and could be adapted to fight assaults on other cells and tissues.

Once artificial chromosomes provide a reliable way of inserting genes in human embryos, we will be tempted to move beyond repairing defective genes or adding beneficial ones. We will want to expand upon existing cellular mechanisms when we see ways of enhancing their functioning. A possible approach to protecting humans from prostate cancer illustrates one such modification.

The strategy requires two genes and their regulatory elements. The first gene is for a poison that will kill the cell. This gene is regulated by a factor that turns it on only in the presence of a hormone that is not found in the body unless it is injected. The second gene has a control sequence that activates it only in cells vulnerable to prostate cancer. The key to the scheme is that the second gene codes for the hormone-dependent factor that regulates the first gene.

This way, the gene for the poison can be turned on only in the prostate glandular cells — where the other gene is active — and only when the hormone trigger is injected. If a man with this genetic modification found he had prostate cancer, he would go to his physician and receive an injection to kill all the cancerous cells.

This is not an outlandish scheme. Each of its key components exists today. Promoters active only in prostate glandular cells could simply be copied. Many cellular poisons are known. All animals have hormone-based gene regulators. Ecdysone, an insect hormone that regulates gene activity in this way, could serve as an injectable activator in humans, because unlike tetracycline, which is also used to regulate genes, ecdysone is unknown in mammals.

Like the AIDS protective described above, this cancer-fighting scheme is a general one. With a different promoter, the technique could be employed in breast, pancreas, or other cancers. Of course, geneticists would no doubt enhance this tool to make it safer and more targeted. They might add extra regulatory controls so that activation of the poison required a cocktail of several drugs, or they might use no poison at all and elicit the cells' "suicide" program (apoptosis) or independently target different cell types. Geneticists might even abandon today's seek-and-destroy paradigm for fighting cancer and adopt emerging target-and-manage concepts, which focus not on killing cancer cells but on interrupting the sequence of cellular changes that allow them to grow and spread.

Such methods are not conceptually complicated and may not prove particularly difficult to put into practice. A critical feature of these germline interventions is that they address adult conditions and act in ways that are specific and of limited duration. They do not greatly exceed the limits of our current understanding. And because the recipient of such modules could always choose not to activate them, so long as they are safe in their inactive state, it matters little if later technologies supersede them or if a recipient later decides he doesn't want them.

Chromosome 47, Version 5.9

The preceding examples give us a glimpse of what mature human germline engineering might look like. At its foundation would be one or more auxiliary chromosomes with hundreds of sites where independent genetic modules could be inserted. These chromosomes would be configured so that a recipient could either pass them on to his or her children or discard them. Various genetic markers would make it easy for geneticists to identify the chromosomes and monitor their operation. Insertion sites might even have special sequences that serve as targets for future somatic additions in adults.

Anything that facilitates yet-to-be-developed adult gene therapies will be of immense value, because germline interventions occur at conception and the elucidation of our genetics will probably be ongoing. A present obstacle to workable somatic interventions is the difficulty of selectively targeting individual tissues. Moreover, in most somatic interventions, regulatory sequences are too big to carry along with an inserted gene. Using germline interventions to preinstall tissue-specific regulatory sequences might allow better control of a future installed gene.

Most modules on artificial chromosomes, at least initially, will probably mimic and extend genes and regulatory elements found in nature, but eventually clever hybrids and new genes may appear. Government and industry no doubt will attempt to ensure quality by adopting standards of testing and design, just as with drug approvals. This is similar to design and quality-assurance programs outside biology and is an intriguing expression of the merging of human and machine. The union is more than a physical cyborgization or a functional fyborgization; it is a convergence of the processes that bring us as well as machines into being and shape our natures. Human conception is shifting from chance to conscious design.

The analogy between artificial chromosomes and computer programs provides a hint of how we may handle our genetic future. Software companies patch faulty software programs and periodically update them, adding new features and incorporating previous patches. Consumers live with the bugs, get patches to correct bad errors, and eventually upgrade to new releases. I wrote this book using Microsoft Word 9.0, my third or fourth version of the program and probably not my last.

Artificial chromosomes may show similar patterns. Imagine that a future father gives his baby daughter chromosome 47, version 2.0, a top-of-the-line model with a dozen therapeutic gene modules. By the time she grows up and has a child of her own, she finds 2.0 downright primitive. Her three-gene anticancer module pales beside the eight-gene cluster of the new version 5.9, which better regulates gene expression, targets additional cancers, and has fewer side effects. The anti-obesity module is pretty much the same in both versions, but 5.9 features a whopping nineteen antivirus modules instead of the four she has and an anti-aging module that can maintain juvenile hormone levels for an extra decade and retain immune function longer too. The daughter may be too sensible to opt for some of the more experimental modules for her son, but she cannot imagine giving him her antique chromosome and forcing him to take the drugs she uses to compensate for its shortcomings. As far as reverting to the pre-therapy, natural state of 23 chromosome pairs, well, only Luddites would do that to their kids.

Intriguing questions arise. What formats will such chromosomes have? Will vendors develop rival designs? Will new versions be compatible with earlier ones? Will uniform standards be enforced? The answers actually may not matter much, because inheritance of these rapidly evolving structures will be so problematic. Few people will want to pass their auxiliary chromosomes on to future generations, and those who choose to do so will likely need to reintroduce the chromosome through *in vitro* fertilization anyway. Simple sex won't suffice, unless the person's partner happens to have an extra chromosome too, and probably one of the same type.

Some biologists believe that scenarios like this are sheer fantasy. They think the complexities of gene-gene interactions would produce too many unintended problems. The nature of reproduction suggests, however, that such genetic crosstalk is far from insurmountable. Despite the many interactions among genes, sexual reproduction creates diverse permutations of gene variants with few adverse effects. Each child gets a new shuffle of the genetic deck. If robust compensatory mechanisms did not buffer the critical aspects of our physiology from the vagaries of gene-gene interactions, only an occasional shuffle would survive, much less thrive. But except for outright errors like broken chromosomes, extra chromosomes, and occasional pairings of deleterious gene variants, most children have genomes that work well.

Future genetic engineers may find that predicting the outcomes of new gene groupings is difficult, but most combinations probably will not lead to serious dysfunctions. The risk will diminish further if, as seems likely for early human engineering, genetic designers model their work on fortuitous combinations of gene variants that already have arisen naturally in lucky individuals. And if problems stemming from such interactions did turn out to be common, the difficulties would be readily apparent in tests on mice and other animals, since comprehensive animal interventions will necessarily precede reliable human ones.

The biggest challenge we will face from germline technology is not from its failure, since that would leave us where we are now. Success is what will tax our wisdom, because that would force us to come to grips with the medical, social, political, and philosophical implications of self-directed human evolution. I have described a simple adjunct to IVF, the injection of artificial chromosomes loaded with genetic supplements. It may prove to be humanity's best hope, and its worst fear.

5

Catching the Wave

> Let us imagine a number of men in chains and all condemned to death, where some are killed each day in the sight of the others, and those who remain see their own fate in that of their fellows and wait their turn . . . This is the picture of the condition of man.
>
> — Blaise Pascal, 1623–1662

The Dying of the Light

A sad irony of life is that brutal decay is the fate in store for each of us lucky enough to live long enough to reach it. Although preimplantation genetic diagnosis may soon enable us to protect our children from various age-related diseases, it will at best only modestly defer their reckoning with Father Time. Genetic manipulation might accomplish more, and in light of our yearnings for immortality, the underlying biology of aging may well be the first germline intervention to truly tempt us.

The possibility of controlling human aging reveals how midlife interventions, using coming biological knowledge, may influence our attitudes about similar embryo manipulations. Moreover, both arenas expose the inadequacies of current medical testing methods for appraising any serious reworking of human biology.

If you could safely add a genetic module to an embryo and thereby give your future child extra decades of healthy life, would you? Whenever I ask this question in public forums, many people say they would, but quite a few say no. Appealing as the idea of lengthening the human lifespan may be, some find it deeply disturbing. We already live long enough, they argue. There are too many people on the planet now, and living beyond a normal lifespan would be selfish. Besides, death is a natural part of life and gives it its meaning.

General concerns about overpopulation and selfishness, however, are not the same as letting go of one's own life. And if death really does give meaning to life, it doesn't follow that a longer life is less meaningful. If the average human lifespan doubled, I suspect we would soon adjust our vision of life to encompass a more expansive terrain. Indeed, before long we would probably view a fatal accident at age 120 as every bit as tragic as one at sixty now seems. After all, a century ago, when the average lifespan in the United States was forty-five, few people would have viewed a grandfather who died of a heart attack at fifty-nine as having his life cut tragically short, but we do now.

A twenty-year-old may still imagine that seventy-five well-lived years are enough, but ask some healthy seventy-five-year-olds busing through Spain for the first time if they are ready to leave this world behind. Precious few will say yes, even if they have come to accept their mortality. They know that one day they may welcome death as their deliverer, but only because they are tired of suffering poor health, have lost their life companions, or are too broken in spirit to keep fighting.

The large markets for face-lifts, hair implants, and Viagra are testimony to the difficulty of accepting the steady erosion of our vitality and youth. If we could hold back time's hand, I suspect many of us would. Vague hopes that life expectancies will rise substantially because of healthier lifestyles and better medicines are wishful thinking. We are pushing our present biological limits. If the age-specific death rates in the United States were to continue to decline at the present pace, average life expectancy still would not reach eighty-five until the year 2180. To stretch the trajectory of human life in a major way will require fundamental breakthroughs in our ability to control the aging process itself.

The present goal of the field of biogerontology is not to conquer aging or extend our natural lifespan, however, but to keep us healthy and shorten our final period of decline. At first such a project sounds admirable, but take this ambition to its logical conclusion and it becomes obvious how out of touch it is with our true aspirations. Imagine that medical science succeeded gloriously in achieving its goal of shortened morbidity. We would retain our youthful vigor for our entire lives and compress our final decline into a painless couple of months, just enough time to say our goodbyes and put our affairs in order.

Such a death, pulling us from active life in our seventies or eighties without the ever-worsening debility that forces us to disengage from the world and confront our inevitable departure, might be far crueler than what we know today. When the time finally came to die, not only would we feel wrenched from our prime, we would leave behind a more gaping hole. Our families would not have been pushed to prepare for their coming loss, nor would they have had the consolation of knowing that we had at least escaped from pain. If the human lifespan proves immutable, of course, better health will sound pretty good. But why not acknowledge that what we really desire is not shorter morbidity at the end of life, but a life that is both healthier *and* longer?

The real question is not what we want, but whether it is too much to hope for. Spending billions to combat stroke, heart disease, cancer, Alzheimer's, and other age-related impairments isn't wrong, but substantially slowing the underlying process of aging might have more impact on these illnesses than today's disease-specific methods ever will. Research into the biology of aging aligns better with our true aspirations for medicine, because it is less likely to extend the time we spend in decrepitude, battling one disease after another. Slowing the aging process has the potential to postpone not only the age-related diseases that blight our final years, but the general age-related decline that parallels them.

Present prospects for retarding or even reversing key aspects of human aging are reasonably good. Today's efforts are not a quixotic reprise of Ponce de León's doomed search for the fountain of youth. Genes shape the patterns of aging in animals. A mouse, a canary, and a bat are warm-blooded and about the same size, but have lifespans of three, thirteen, and fifty years, respectively. One day we may modify our genetics to extend our lifespan. After all, with only modest changes, evolution did as much for our chimpanzee ancestors: though only 1 or 2 percent of the bases in our genomes differ from theirs, we live some 50 percent longer than chimps do.

As we unravel human genetics, we will come to understand the mechanisms by which we age. No one yet knows whether we will be able to intervene genetically or pharmacologically to retard the process, but tantalizing clues suggest that both may be possible. Scientists have more than doubled fruit fly and roundworm lifespans by changing single genes. And they have increased mouse lifespans some 40 percent using dietary restrictions of caloric intake.

These two modes of intervention have different implications, however, since we would use one on ourselves and the other on our unborn children. To understand how interventions to retard aging might affect humanity, we must understand their likely power, the degree to which they will be genetic, and whether they will require germline selection and manipulation of embryos. Not surprisingly, these same considerations will apply to many other potential germline interventions.

The lifespan profiles for different species are strictly a matter of genetics, but the length of your own life will depend on its details and vicissitudes: whether you live under constant stress, smoke and drink, eat vegetables, exercise, wear a seat belt, or are lucky. A comparison of the relative longevity of thousands of pairs of twins born in Denmark in the late 1800s suggested that genetics accounted for only some 25 percent of the variation in their age of death. These studies looked at few of the very old, however; early studies of Sardinian centenarians suggest a greater genetic contribution to extreme longevity. Furthermore, heritability may be higher for those living today in the more predictable environments of the developed world. Mothers rarely die in childbirth; we usually survive infectious diseases; a social safety net protects us; the killing fields of World War I are gone, along with the deadly 1918 flu pandemic.

Genetic analysis is not an especially useful tool for predicting how long a particular individual will live. Some mouse strains are particularly long-lived, for example, which is largely due to genetics, but the length of their individual lives varies greatly. A cohort of genetically identical mice, caged individually and treated virtually identically, does not die within a narrow band of time. Roy Walford, a leading authority on the biology of aging and the researcher who pioneered work on rodent dietary restriction, has done numerous experiments comparing mouse lifespans with and without restricted food intake. In a typical experiment, the first deaths in his control group — the mice with ample feeding — occurred at only 16 months of age, most of the animals died between 32 and 38 months, and the last mouse keeled over at the ripe old age of 42 months. Given that they all had virtually the same genes and the same environment, that is quite a spread.

In the experimental group (same genetics, same handling, except for a 40 percent reduction in calories), however, the first death was not until 32 months, most of the mice died between 47 and 55 months, and the last one expired at 57 months. The variation is still large, but average and maximum longevity have risen by more than a third. Such experiments leave no question that retarding adult mammalian aging without altering genes is possible, even if we don't yet understand what is going on physiologically. And this is not unique to mice; similar effects are now being found with primates.

But self-starvation is no picnic. Only a handful of people have taken these results to heart and drastically restricted their own diets, and the Internet user group devoted to this lifestyle is reputedly one of the grumpiest around. Some researchers hope to circumvent this unpleasant regime by understanding the mechanisms of caloric restriction and developing genetic or drug interventions that emulate the underlying physiological changes that make it work. Richard Weindruch and his colleagues at the University of Wisconsin, for example, are using DNA chips to screen thousands of genes in normal and calorie-restricted mice in the hope that the differences in gene expression they find will lead them to the precise metabolic pathways involved in postponed aging.

Whether such research will produce practical clinical interventions in adults or genetic interventions in embryos remains to be seen. In the next decade, as we identify the constellations of genes that add to human life expectancy, we may find that even gene variants common among the longer-lived may add no more than a decade or so to average individual life expectancy and do nothing for such people as Jeanne Calment, who died in France in 1997 at the age of 122, the oldest human being. To achieve a doubling of the human lifespan — a concept often tossed around in the media — will require something more radical: an understanding of the genes associated with aging and a deliberate alteration of them to inhibit, circumvent, or compensate for normal aging mechanisms. This will probably, but not necessarily, require germline engineering.

Dreams and Pipe Dreams

At present, no one knows how much the human lifespan might be extended — no upper limit has yet been identified. As we learn more about the genetics, physiology, and biochemistry of aging, increasingly effective ways may emerge to control it. A generation ago, few researchers imagined that altering human genetics to retard aging might become practical. Now, with the large, genetically engineered enhancements of lifespan in short-lived invertebrates and the explosion in our capacity to modify mammalian genetics, the possibility seems real. Yet the notion of reversing aging has always been a pipe dream. Such a hope seems like mere denial. Surely no reputable scientist would seriously suggest that a frail, hobbling octogenarian might one day undergo some treatment and come out of it seemingly youthful or middle-aged.

Or so I thought. In June 2000, Aubrey de Grey, a biologist at the University of Cambridge in England, asked me to help him organize a roundtable of mainstream researchers who had been considering the challenge of reversing various manifestations of aging. The meeting took place that October, hosted by Bruce Ames, a distinguished biochemist at the University of California at Berkeley. Ames is not a fringe figure. He is a member of the National Academy of Sciences, the winner of a dozen prizes, including the 1998 National Medal of Science, and a prolific researcher whose more than four hundred papers include extensive work on the role of micronutrients in cancer and aging.

The meeting examined possible strategies for restoring muscle mass and function to youthful levels, rebuilding synaptic connections in the brain, breaking down the collagen cross-linking that reduces limb flexibility, restoring youthful hormone levels, eliminating faulty cells that disrupt tissue, and even rebuilding dysfunctional mitochondria, the organelles that power all our cells. Clearly, this was an ambitious agenda, but the scientists concluded that a reasonable goal would be to reverse key aspects of aging in mice within a decade as a prelude to human interventions. What most surprised me as I listened to the presentations at this small meeting was that we were discussing such things at all.

The participants concluded that reversing aging might be easier if researchers stepped back from present efforts to tweak existing metabolic pathways and instead sought novel ways of manipulating the essentials of the process. To that end, the participants drew up a list of bioengineering projects that would attack key aspects of aging and open a new course for aging research.

Although the suggested approaches were highly speculative, they were firmly rooted in existing research. For example, when enough genetic mutations or other insults compromise a cell, it typically assumes a "senescent" form and dysfunctionally churns out chemicals that degrade its immediate surroundings. As senescent cells build up over the years, the internal environment of the tissue deteriorates, healthy intercellular signaling degrades, tissue function diminishes, and problems feed on one another to produce aging or, in some cases, a cancer. Such senescent cells may be few in number, but Judith Campesi, a leading advocate of this theory of aging, suspects that their disruptive influence may be disproportionately large. To try to combat this problem, scientists could modify the workings of cells to improve their repair mechanisms and reduce the damages that push them toward senescence. But such a project would be an immense undertaking, and alterations to basic cellular functions could easily have unforeseen consequences.

A more direct approach would be to accept that the body is using senescence to try to shut down damaged cells, and help the body out by deleting them entirely. To accomplish this, the genome would need an added "cell-death" gene, programmed to turn on in cells that began to show telltale signs of senescence, or perhaps the cells' own built-in capacity to self-destruct, called apoptosis, could be triggered. Whether such a strategy would actually slow aging is unclear, but Campesi is trying to find out by modifying the germlines of mice.

If she can slow aging in these animals by deleting senescent cells as they arise, she might reverse aging as well, by sweeping them from adult tissue. To test this, she would block the cell-death construct by putting it under the control of a tetracycline switch, let senescent cells build up until the mice reach late adulthood, then use a dose of tetracycline to wipe out the cells. Whether the tissue would revert to its prior healthy state or remain compromised would be a strong indication of the technology's potential usefulness in combating aging in adults. Even if reversed aging occurred in mice, however, adult therapy in people would face a major hurdle, one that does not faze germline technology: how to improve today's somatic-therapy techniques enough to put the genes into adult cells.

The social and psychological implications of a germline therapy that can arrest some significant aspect of aging will hinge on whether geneticists can adapt it for adult use, but even an intervention in mice might have important immediate consequences. Such a success might bring a critical shift in public perception by undermining today's sharp distinction between aging and age-related diseases such as cancer, diabetes, arthritis, and heart disease — manifestations of aging that are present to some degree in most older adults. In fact, we might start to see aging not simply as *a* disease, but as *the* disease. It affects everyone, it cripples, it kills, it is brutal, and suddenly it would be seen as potentially treatable.

A similar calculus inspired President Nixon's 1971 declaration of war on cancer. That broadly defined effort's massive infusion of funds into basic biomedical research laid the foundation for many of the advances being achieved today. Now, with our growing armamentarium from genomics, the time may be arriving not only to realize serious progress against cancer, but to begin a broader campaign against disease and debility — a war on aging.

The Challenges Ahead

Progress against aging provides an instructive example of the profound changes awaiting us as we begin to modify our biology. If the human lifespan doubled, the changed trajectory of human life would transform our institutions and our lives. Virtually every aspect of society would shift: patterns of education, work, and marriage, relationships between parents and children, attitudes about social investment and responsibility, the flow of wealth and opportunity from one generation to the next. Our vision of life's possibilities, our assessment of our risks and rewards, and our goals and aspirations would all change.

Seeing the immensity of the consequences of postponed human aging is easier than figuring out precisely what they would be. The specifics depend too much on the nature and relative timing of potential interventions. Aging is multifaceted, so therapies that emerge will likely be an amalgam of behavioral, dietary, genetic, and pharmacological elements, with differing efficacies against various aspects of the process.

A future coupling of youthful bodies and ossifying brains is not a pretty picture and differs from a future in which cancer is still a major problem but muscle strength, bone density, and cognitive function are restored. Furthermore, our attitudes about such treatments will be strongly influenced by whether they have unpleasant side effects, need frequent repetition, are arduous, take long to act, must begin before some specific age, and reverse aging or simply stall it.

The social implications of even a perfect adult intervention would also depend on whether it required specialized lab work that essentially precluded widespread public use. If so, class conflict over access to the technology might follow; if not, enormous pressures might develop to make the treatment available to everyone. Ultimately, the largest social adjustments to such an intervention would probably not relate to oft-discussed topics such as stresses on social security, the aging of the workforce, or overpopulation. The hardest issues would be intergenerational differences in wealth and power, greater gulfs between nations and peoples resulting from unequal access, and the need to reorient our biological drives to match the elongated arc of human life.

People often assume that social changes resulting from germline interventions to extend the human lifespan would be slow in coming. Consider the following wildly exuberant scenario. Medical researchers develop a germline intervention to triple the human lifespan. They create it in a mere twenty years, together with all the subsidiary technologies needed to make it widely available. Everyone in the world embraces the intervention immediately, and in only a decade we erect the clinical infrastructure to enhance every child born anywhere in the world.

Even with these absurd assumptions, no shift in the age distribution of our population would occur until nearly a century from now, when the leading cohort of these enhanced humans would reach the age when debility and death would otherwise cut their ranks. Such analysis is misleading, however, because it ignores the social, political, and personal changes that would precede this demographic shift.

Any strong signal that the human lifespan is mutable and that we might significantly postpone physical decay and death would have immediate consequences. Moreover, an indication of the feasibility of the effort to control aging may well emerge within the next ten years, so it is worth considering what might happen if signs are encouraging.

Imagine that reports of a successful germline intervention postponing aging in mice hit the news this weekend. The findings, slated for publication in next Friday's *Science,* a renowned scientific journal, are solid and detail a 50 percent extension of mouse longevity when researchers activated a cluster of genes in middle-aged animals. The article tells how researchers inserted a half-dozen genes in a long-lived strain of mice that normally lives three and a half years. When the genes were activated in two-and-a-half-year-olds, the mice grew more energetic. A few died during the next year, but most lived one to two years longer than is typical, and one is still alive at the age of six. This Methuselah of mousedom is the oldest mouse in history. The researchers cautiously point out that the intervention lies well beyond the current reach of human somatic therapy and that any germline intervention in humans would be far too risky to contemplate.

Such a report would create a psychological tremor far greater than the announcement of Dolly's birth. Human cloning may one day tempt the occasional eccentric or those with unusual family histories or reproductive problems, but most people don't find the idea seductive. Not so with recapturing or preserving youth.

Initially, of course, skepticism about the likelihood of any near-term human application would abound. Bioethicists would point out that only generations of testing could ensure safety, that side effects could crop up decades down the line, and that parents should not be allowed to engage in such dangerous experimentation on their future children. Theologians would oppose on moral and philosophical grounds the notion of such tampering, even if it could be proven safe. Environmentalists would bemoan the selfishness of the idea in a world already overpopulated. And many would be intrigued but cautious, figuring there had to be a catch.

As with Dolly, a researcher or two would challenge the results and put forward good reasons why they might not be true. Governments no doubt would convene commissions of scientists, ethicists, and spiritual leaders to make policy recommendations, and most of these voices would emphasize the need for broad public discussion of the uncertain consequences of the technology. But as scientists repeated and expanded the research, the realization would start to sink in that the genomics revolution would soon be doing much more than telling us our cancer risks, curing genetic diseases, or personalizing drugs. Gradually, our agonizing about playing God and our worries about longer lifespans would give way to a new chorus: "When can I get a pill?"

A successful germline intervention that significantly extended mouse longevity would ignite a huge effort not toward germline interventions in human embryos, but toward clinical interventions in adults. As Aubrey de Grey, the tall, full-bearded, rather eccentric British theorist on aging, who organized the roundtable on reversing age-related decline, said to me, "Germline manipulations on future generations don't interest me. I already *have* aging." So do we all. Researchers are using germline techniques as a tool for understanding aging and figuring out how to combat it directly in adults, not as a prelude to human germline manipulation, though this may prove easier.

Any drug that retards key aspects of aging would have sales that dwarf today's disease-specific blockbusters. Virtually every adult might take anti-aging medications for life — and a dollar a day from everyone older than forty-five is $30 billion a year in the United States alone. Given such incentives, biomedical science is likely to come up with adult therapies for at least some facets of aging, but ultimately, germline technology may prove a more potent tool for so intrinsic a part of our biology.

A New World of Testing

The validation and safety issues surrounding both genetic and pharmaceutical interventions to retard aging will be immense with whatever biological enhancements come on stream. Existing oversight practices will be of little use in evaluating these future procedures. The arrival of anti-aging drugs could force a complete reworking of clinical testing methods, if not the entire drug approval process.

Judging by the wide use of illicit drugs for recreation and performance enhancement, many people would be unwilling to wait long for the Food and Drug Administration to approve medicines that might peel away the years. Such drugs, after all, would be irresistible. If they could tune up our metabolism to make us feel younger and more vital, they would be so psychologically addictive that a black market would spring up overnight if we outlawed them. No regulatory agency could hope to block such products. And how would the state punish people caught taking these medicines?

Throw them in prison? Bar them from sporting events for seniors? Lock them in a special camp and make them smoke cigarettes, gulp french fries, and eat pastries until they look and feel their age?

Human-growth-hormone supplementation offers a revealing glimpse of how eagerly people will embrace anti-aging drugs, especially when hyped by advertising. Here are banners from three different Web sites: "Grow Young with Human Growth Hormone!" "Aging — Don't Accept It, Turn Back the Clock Now!" "Unleash Your Youth with Professional-Strength Human Growth Hormone!" These sites are not the cause but the response to our yearnings.

The problems with human growth hormone are a troubling preview of the future. The drug is in wide use, true believers and charlatans alike are promoting it, and while some anecdotal information suggests its value, other data are worrisome. No one has yet done the expensive clinical studies needed to clarify the situation, and no one is likely to. The pharmaceutical industry has no incentive, because the patents on human growth hormone have expired. Individual clinics would be risking their livelihoods — their mission is to treat patients, not to do research. Academic scientists are more interested in mainstream questions and could not get the necessary grants for this type of study even if they wanted to.

The best way of escaping this impasse might be to provide existing users of growth hormone with information about the risks they are taking and enroll them in a voluntary trial to capture and analyze the results of their self-treatment. This would not be nearly as useful as a large, highly controlled clinical trial, but the technique would be fast, cheap, practical, and self-adjusting. It would enable users to better evaluate their risks, let researchers resolve unanswered questions about long-term use, and allow both raw information and evolving treatment advisories to be fed back to patients regularly.

Ultimately, this might be the safest route for future anti-aging drugs as well, because testing them in an acceptable amount of time using traditional studies will be virtually impossible. To evaluate a medicine's promise of retarded aging would require not the seven or eight years now typical for drug approvals, but decades. Furthermore, the sizable long-term trials needed to tease out subtle side effects would be extremely expensive, and maintaining the blind controls — in which some participants unknowingly take placebos — would be nearly impossible. Subjects are not going to take some unknown drug faithfully for a decade, especially if they are noticing no benefit.

In the years ahead we will need novel testing approaches for drug and genetic interventions, so regulators should begin experimenting with more flexible methods now, to find and correct their weaknesses. Animal testing and early-phase human trials will take us part of the way toward establishing safety and demonstrating efficacy, but designing human testing that is flexible enough to keep people involved, rigorous enough to provide meaningful information, and worthwhile enough to persuade people to participate is also necessary. Ad hoc trials such as the one I suggested for human growth hormone depart from present policy but could harness the experiences of those who will be experimenting with these drugs anyway. Moreover, the oversight and data sharing would help protect them from misinformation and protect others by gathering the data needed to evaluate and refine the treatments. The technique is reminiscent of the valuable informal networks that appeared in the 1980s among those with AIDS.

Research in which the number and identities of trial participants, as well as their medication regimes, are in continual flux would be of limited value without modern technology to gather and analyze the data. But we now have the ability to take advantage of such protocols. Today, massive self-experimentation with medications and diets is going on throughout society, and we are making almost no effort to collect and use the information. It is a great waste. The world is not neat and clean: people are using performance-enhancing drugs, questionable dietary supplements, and narcotics. We live in the Wild West when it comes to alternative medicines and neutraceuticals (plant extracts that the U.S. Food and Drug Administration does not regulate). Many people diagnose themselves, see multiple practitioners, pick and choose their medicines and supplements, obtain their information from friends or the Internet, and use homeopathy and other unlikely treatments.

Whether we like it or not, we are ever more engaged in a collective global discussion about our health and are groping for ways to improve it. Let us acknowledge this and take advantage of it, instead of relying solely on today's gold-standard clinical tests. The coming regulatory challenges with pharmaceutical interventions will be a wonderful training run for germline enhancements, because these deeper biological manipulations will be even harder to evaluate.

Into the Germline

We have seen that aggressive attempts to develop human anti-aging treatments are bound to follow any germline experiments that cause mice to live longer, healthier lives. We have also looked at what might happen if adult treatments become available. But comprehensive adult treatments may never appear. The push by medical science to find ways of curbing aging in adults might fail. The only way to control key aspects of aging might be through germline manipulations that adjust the genes of a person's every cell.

The longer the delay between the realization of successful germline procedures in mice and the arrival of adult anti-aging interventions in humans, the more parents will be tempted to alter the genes of their unborn children. In fact, this may occur even if adequate adult interventions are found, because such treatments, while useful, may not seem as good as germline ones. In either case, parents might see the single-cell embryo as a momentary opportunity to give their child gifts otherwise lost to him or her forever. Because aging is the most predictable health problem we face, it will always be a nearly ideal candidate for germline preventives.

After artificial chromosome technology matures enough for routine research use, germline successes in mice will inevitably lead toward equivalent human therapies. Not only will researchers refine human protocols through mouse studies, they will find that their successes with those models will strongly influence public perceptions about and funding for human work. Any serious progress in manipulating aging in mice will go a long way toward bringing about a general acceptance of human germline modification, because it will demonstrate the technology so tangibly.

The more we succeed in modifying our biology and that of other animals, the more we will see it as something malleable that we can adjust and improve, and the more we will come to assess germline therapies on the basis of risk and reward rather than philosophical meaning. If the effort to extend human vitality and lifespan includes germline intervention — and anything else seems unlikely, given its potential and the many obstacles facing somatic therapy — then broad acceptance of germline technology for other purposes will be on the way.

Robust germline procedures to combat aging would demonstrate the devices, such as artificial chromosomes and gene-expression controls, that underlie the technology. The interventions would demystify the technology by offering walking proof of its potential. And the effort would establish methods for ensuring safety.

Again, the issue of safety is critical, because it is so common an argument against applying germline procedures to humans. As with any anti-aging interventions, clinicians will only be able to partially test germline procedures before using them in humans. But this is not unusual in medicine. In 1991 when Andre Van Steirteghem, at the Center for Reproductive Medicine in Brussels, used intercytoplasmic sperm injection (ICSI) to overcome severe male infertility, he had only animal experiments to guide him. There was no certainty that ICSI would work. Much can go awry when you chop the tail off a sperm, suck it into a pipette, and inject it through the punctured wall of an unfertilized egg, as he did. But ICSI did work. And as a consequence, male infertility today is rarely an obstacle to *in vitro* fertilization — which, of course, also was untested in humans when it was first used. Before ICSI, a man needed millions of actively swimming sperm to father a child. Now a few nonmotile laggards suffice.

No one will be able to confirm the success of germline procedures, particularly against aging, until many decades after they occur. Until then, confidence in their safety and efficacy will have to come from clinical procedures on older adults or from germline interventions in short-lived laboratory animals. The mouse, with its three-year lifespan, will likely remain the mainstay of aging research.

Given the unavoidable questions about safety, early germline interventions to slow human aging will probably employ a module of genes that are blocked until the recipient decides to activate them. Such conditional expression might make this procedure even less risky than taking medication. Both methods could be stopped and started at will, but the genetic intervention, unlike the pharmaceutical one, would act in a highly localized way, touching only certain intended cells.

Such delayed activation of germline modules would have the side effect of making primate studies particularly valuable, because they could run in parallel with human use rather than precede it. The shortest-lived primate that researchers might use is the lemur, which lives some fifteen years. If identical gene modules were installed in human and lemur embryos, for instance, enough time would pass while a person was young and his or her module was blocked to see how the lemur module worked. When the time came to decide whether to activate the module, a person would have results from primates and generations of mice. As long as the shutdown of the module was foolproof, the scheme would be safe.

The biology of aging is still a peripheral field, but once we perceive that aging is not simply an inevitable process of decline, research will explode, with greater resources, new talent, and a focus on achieving clinical interventions. Society will begin to consider how much to spend on efforts to tame aging, how to ration the therapies that become available, and how to make various social adjustments. The initiation of a plausible effort to extend the human lifespan would spark other debates as well: about the wealthy buying their children the added life that is unaffordable to others, about the danger of divisions between those who can afford such procedures and those who cannot. These are the very issues society will face with other biological enhancements.

Some will want to block these interventions or slow their development, but others will want to step up the pace. A longer and healthier life is a near-universal goal, so more people will embrace this biological enhancement than any other. People who don't care about an extra few IQ points for themselves or their kids may care a great deal about added years of vitality. Because anti-aging procedures may turn out to be the first human enhancement with widespread appeal, their course may shape our attitudes about biological enhancement in general, including germline modifications.

Whatever our attitudes about biological enhancement, I suspect that most of us would rather be among the first to live an extended lifespan than among the last to live a "natural" one. Yet the idea of striving to extend our lives is somehow discomforting. We celebrate the nobility of self-sacrifice and the heroism of risking death for the common good. We do not applaud those reaching for longevity. Their self-serving actions evoke images of cowards on the deck of the *Titanic,* pushing aside women and children to clamber into lifeboats, or hypochondriacs counting their every vitamin and avoiding anyone with a cough, or, yes, even vampires sucking the blood of others to buy immortality. It is easy to recoil from those who practice caloric restriction and starve themselves in the pursuit of added years. The travails of their quest are reminiscent of religious ascetics, but their goal seems unworthy and narcissistic.

We should look more closely at the source of our repugnance, however. Today those grasping for longer life seem to be relinquishing life's richness and focusing solely on themselves and their survival. Small wonder at the disdain they sometimes provoke. But real breakthroughs in the biology of aging would change this. We might not be willing to starve ourselves to buy a few years, but surely we would take a pill. This would be neither selfish nor self-absorbed; it would be common sense.

If a germline procedure could double our lifespan, the egotistical pursuit of more years would not tarnish the beneficiaries any more than antibiotics and vaccines tarnish us. Our children's extra years would come as naturally as ours have. Nor would our longer-lived children be thought selfish because they require additional resources in their added years. If we judged lives this way, we would not smile approvingly at those who reach their hundredth birthday in good health.

Longer and more vital lifespans would not only have personal consequences, they would also enrich society. The family cycle of pain and disruption from old age and death is clear to everyone, but economists try to be more specific. William Nordhaus, an economist at Yale University, estimates that half the increase in the standard of living in the United States during the past century is due to the rise in longevity that has lengthened our active lives.

The benefits to society of extending our vital years are as clear as the burdens of prolonging our decrepitude. We require decades of education and experience to learn to handle ourselves effectively in the world, but we tire and fade all too quickly. Added years of health would lessen this drain. If youth is wasted on the young, then why not see what the old can do with it? The result would undoubtedly be good for the individual, the family, and society.

The implications of added longevity are not easy to fathom, and those of other germline manipulations will be no easier. As we consider the many possible enhancements attending our emerging capacity to rework our biology, we must not be too quick to judge them.

6

Targets of Design

Nature is often hidden, sometimes overcome, seldom extinguished.

— Francis Bacon (1581–1626)

WE CANNOT say much about the challenges that will accompany the first steps of human self-design until we examine the specific biological modifications that might intrigue us when the technology arrives. Preventing disease, improving health, and increasing longevity will surely have strong appeal. But will increased intelligence, better memory, and greater strength also inspire us? And will parents want blue eyes and blond hair for their children, or even such quirky traits as skin that glows in the dark?

The answers may be right in front of us, in the choices we make today when we modify and enhance ourselves through aesthetic surgery, prosthetic implants, and electronic and mechanical extensions. These present choices are a preview of the deeper ones we will face in the not too distant future, because they reveal the cultural and biological desires that shape our preferences.

We must, of course, remain firmly grounded in the real world. We might like the idea of flying like Superman, but it isn't going to happen. Nor could we genetically determine our native tongue, because language acquisition is independent of genetics. We must restrict our focus to the zone where desire and feasibility intersect.

Underlying any attempt to discern the attributes parents might one day shape in their children using genetic selection and modification is the question of the degree to which genes influence particular traits and how complex their influences are.

To argue in general terms for the importance of genetics is hardly necessary today, given the frequent headlines proclaiming the discovery of genes for obesity, risk taking, sleeping patterns, seizures, poor eyesight, and other attributes. Only after reading the accompanying articles do we see that the identified genes account for just a small percentage of a trait's incidence or that a particular study was small and preliminary. Reacting to the headlines, the public tends to inflate rather than understate the power of our genes, imagining they will lead us to miracle cures for most diseases.

People also seem willing to accept that their genes may have fashioned subtle aspects of their personalities — their conscientiousness, shyness, friendliness, intelligence, or tough-mindedness. This represents a big change from twenty-five years ago, when Edward O. Wilson, a professor at Harvard University, provoked a storm of controversy by arguing in his book *Sociobiology* that our evolutionary history and biology influence cultural practices such as marriage and religion. Such ideas, though they still upset some people, are commonplace today. But in 1975, they so ran up against accepted dogma about social engineering and individual plasticity that they elicited harsh criticism. The well-known essayist and evolutionary biologist Stephen Jay Gould, for instance, attacked Wilson aggressively, asserting that such theories as his had served as a foundation for Hitler's eugenic policies.

Given the Nazi legacy, many people would like to view children as lumps of clay shaped by proper rearing and a beneficial environment. But any parent without blinders knows that kids have their own natures and personalities, and that many of their tendencies are present at birth. Some babies are fussy and irritable, some calm, some inquisitive. The crucial question is not whether genetic influences exist, but how large they are, how reliably they translate into adult personality traits in the face of differing environmental influences, and the extent to which we can modify their influences.

Our challenge is to figure out why someone ends up shy rather than outgoing, optimistic rather than pessimistic, bubbly rather than moody, or prefers mornings to evenings. This might tell us how different a person would have turned out had he or she grown up in a different family or a different place, or been treated differently.

Twin Studies and Heritability

In the 1970s no definitive answers to these questions existed, but well before genomic research began to link specific genes to specific diseases, researchers started using a variety of twin studies to measure the extent to which our genes might explain the variations in human traits. To get a handle on the relative influence of nature (our genetics) and nurture (our environment and experience), researchers compare twins raised in different families. The Minnesota Study of Twins Reared Apart, begun in 1979 by Thomas Bouchard at the University of Minnesota, is the most famous such research. It has extensively examined about 125 pairs of reared-apart twins, including some 60 pairs of identical twins, putting each twin through more than forty hours of detailed psychological and physiological assessments.

Because identical genes but different environments helped shape the identical twins who were reared apart, traits with largely genetic origins are more similar between them than traits influenced primarily by the environment. Such studies are difficult and must be interpreted very carefully, but until we have the technology to screen large numbers of people and analyze their genes using DNA chips, the approach is likely to remain the best way of determining the relative contributions of environment and genetics to specific human traits. Only those traits that vary primarily because of genetic differences, of course, are candidates for embryo manipulation and selection.

Identical twins reared together provide information too. They not only have the same genes, they had the same parents, grew up in the same neighborhood, had the same socioeconomic status, and shared many other environmental influences. Differences between them reveal the influence of factors that are unique to each of us: becoming ill, being punished, winning a competition, having a special teacher or friend.

Comparing fraternal twins, whether reared apart or together, adds yet more data. With only half the genetic relatedness of identical twins, these siblings show less similarity for attributes that are largely genetic. Finally, there are so-called pseudo-twins, unrelated children of the same age who have been raised together. Their similarities point to attributes shaped primarily by shared environmental factors.

Researchers have run hundreds of studies in numerous countries and examined various traits in various types of twins. The results have been remarkably consistent. Genetic factors generally account for between 35 and 75 percent of the variation among people in traits we think of as significant. Environmental influences and random factors that are unique to each individual account for most of the rest, whereas environmental influences shared by an entire family matter little, at least within the range of environments encountered by a typical child growing up in the developed world.

I have been speaking of genes and environment as though they are independent, but this is not the case. Genes not only affect our minds and bodies directly by shaping our biology, they also do so indirectly, by influencing the environment we experience. A child who excels at sports is more likely to gravitate toward athletic activities, just as one who loves to read philosophy might choose more intellectual pursuits. Both children would be selecting their environments. This happens in less overt ways as well. A reclusive, rigid child almost certainly elicits different treatment from those around him or her than one who is gregarious and easygoing. Thus, self-reinforcing feedback comes into play: our biological predispositions shape our environment, which in turn reinforces our predispositions. Some of the spread that exists in estimates of the heritability of IQ, for instance, may arise because of the dissimilar ages of the subjects in different studies. By late adolescence, twins tend to be closer in IQ than they were in childhood, which may be because of their growing power to align their activities with their underlying predispositions. Similar results show up with qualities such as antisocial behavior.

IQ provides a good example of the difficulties we face in teasing apart the relative contributions of environmental and genetic factors. Many studies have looked at the heritability of intelligence, and although some of them have been challenged as flawed or even fraudulent, the work typically shows that IQ is anywhere from 45 to 75 percent heritable. Moreover, some studies conclude that adopted children living together show no more correlation in their IQs than other unrelated individuals. The households covered in these studies do not include extremes of poverty and environmental disadvantage, but absent this, the influence of living in one home rather than another is small.

Not surprisingly, studies of the biology of intelligence are highly controversial. While almost no one claims that IQ accurately gauges all the dimensions of human talent, people who score higher on these tests clearly tend to apprehend, scan, retrieve, and respond to stimuli more quickly than those with lower scores. Moreover, IQ is one of the most useful predictors of school performance, years of education, job performance, and income. Such correlations are statistical, of course, and cannot tell us what will happen to a particular individual. Some geniuses are too dysfunctional even to hold down a job. But these correlations with IQ are real, they are significant, and they may grow stronger as our society becomes more complex and technology intensive. Deciphering, reading, and manipulating the genes involved in human intelligence are going to be thorny political issues, to say the least.

Long before the genomics revolution, stories of uncanny similarities between identical twins were contributing to the idea that our genes determine much about our personalities. The most publicized such twins were probably the so-called Jim twins, James Springer and James Lewis, whose story broke in 1979 in the *Minneapolis Tribune* and inspired Thomas Bouchard to launch his famous twin study.

According to the article, separate families in Ohio adopted each twin a few weeks after their birth in 1939, and the two had no contact for almost forty years. They each had a first wife named Linda and a second wife named Betty. Each had named his first child James and had given him the middle name Allan (or Alan). Each had named a childhood dog Toy. Each enjoyed carpentry and mechanical drawing. Each was six feet tall and weighed 180 pounds. Each had been a part-time law enforcement officer. Each liked Miller Lite beer and chain-smoked Salems. Strangest of all, each had once driven his family in a light blue Chevy to vacation along the same three-block strip of beach in Florida.

Later tests and interviews by journalists revealed even more concordances. Each had a tree in his back yard with a white bench wrapped around it. Each had an elaborate workshop and made miniature furniture. Each was a fan of stock-car racing and hated baseball. Each scattered love notes to his wife around the house. Each was an anxious sleeper, ground his teeth at night, bit his nails to the quick, and had been plagued by migraines since his teenage years. Each had high blood pressure. Each had "lazy eye" in the same eye. Each had hemorrhoids. Each had had a vasectomy. And their scores on tests measuring sociability, flexibility, tolerance, conformity, and self-control were very similar.

An occasional equivalence would not surprise us. Even in random conversation, we sometimes stumble on such oddities with friends. But the above list is overwhelming and perplexing. The items seem too numerous and wide-ranging to be merely coincidental, yet some are so specific that we can find no other explanation, except perhaps outright fraud. But keep in mind that these correspondences are the product of a thorough probing of many possibilities, each no doubt substantially narrowed by the twins' overlapping predispositions. The researchers recorded some fifteen thousand such facts on each sibling they studied.

That each Jim, as a child, named his dog Toy might be a manifestation of both chance and similarities in personality. We can guess that neither Jim was adventurous with names, since they both called their sons after themselves. Perhaps in rural Ohio just a few dozen puppy names were popular among young boys then. Explaining a preference for Salems or Miller beer doesn't require much of a reach: the number of major brands is limited. Similarly, many people hate baseball and like stock-car racing. If each of the thousands of items recorded had even a 1-in-100 chance of matching, many intriguing equivalences would show up for any two people, and many more for two with very similar predispositions. Still, the strange detailed matches that grab our attention in stories about twins are not fraudulent and arise much more frequently in identical than in fraternal twins, which is what we might expect given the interplay of strongly shared predispositions and enough details to multiply coincidences.

Genes tell us much about ourselves, but they only channel our fates, they do not engrave them indelibly. When one identical twin gets Alzheimer's, the other has only a 50 percent chance of being afflicted. But 50 percent is a huge chance if you are such a twin. A good friend had an identical twin brother who began to show signs of Alzheimer's at age fifty-eight and within a decade was unable even to recognize members of his family. When my friend's sister also began to show signs of the illness, he'd had enough of living in the shadow of the disease and, at seventy-one, went to the hospital for a brain scan that would reveal incipient Alzheimer's. For three days he anxiously awaited the verdict. He was lucky: all was normal. No one has an explanation, but you can bet he will continue the anti-inflammatory drugs he has been taking for twenty years for an unrelated condition. His doctor claims he no longer needs them, but maybe they made a difference. A few early reports are now suggesting the medication may inhibit Alzheimer's.

Many physical attributes are highly heritable. We see signs of it in families all around us. Twin studies show that given adequate nutrition, genetics can explain some 80 percent of people's variation in height, 70 percent of it in weight, and 60 percent of the variation in blood pressure. The same strong genetic connection also obtains for mental disorders: autism is reported as 90 percent heritable, schizophrenia 50 percent, bipolar disorder more than 70 percent. The specific percentages shift somewhat from study to study, but the message is clear: in spite of the various personal life experiences that bring on these conditions and shape their manifestations, genetics determines a large portion of our susceptibility. The series of losses that trigger a descent into melancholic depression and suicide in one person do not do so in another with different genes.

The power of our genes to mold us does not stop with our physical attributes and disease susceptibilities. Our genes are the most important single factor in determining great swaths of "normal" personality too. Ambiguities exist in measuring such traits using standard self-reported personality tests, but results are reasonably consistent, with scores being about 90 percent reproducible when tests are repeated on the same person. The big picture from these studies is unmistakable: our genetic makeup strongly influences the defining aspects of personality and temperament. Studies report that our genes account for anywhere from 40 to 60 percent of the variation in personality among us. This includes our level of extroversion and self-involvement, our emotional stability and reaction to stress, our conformity and dependability, our friendliness and likableness, and our general openness and curiosity. Even whether a person says that religion is important in his or her life is about 50 percent heritable.

Remember, however, that heritability is not absolute; it refers only to relative genetic influences within a particular range of environments. Alter the environment beyond those bounds and a trait hitherto unresponsive to environmental change might shift significantly. Starve a child who would otherwise be tall, and he or she will be shorter regardless of genetics. Encourage a very shy child in just the right way, and he or she may become more outgoing. The heritabilities measured in twin studies are valid only for the range of environments where the studies took place. This does not invalidate the use of these results for evaluating a trait's potential for genetic selection, though, because any children chosen on this basis would presumably grow up in a similarly protected environment.

Engineering Human Embryos

Twin studies demonstrate the importance of our genes but tell us neither how these bearers of life's possibilities achieve their effects nor how difficult deciphering and manipulating their influences will be. That we are tall or bright or outgoing or at high risk of cancer because of our genes does not mean that changing or selecting these attributes will be feasible.

A crude gauge of the complexity of modifying a trait in a directed fashion by altering or selecting particular gene variants is the number of genes associated with the trait. Changing a single gene in animals is routine today, even if it cannot be done as safely and reliably as human interventions would require. Manipulating a dozen genes will be much harder, but might be possible in a few decades. If hundreds of genes are involved, however, scientists will probably be unable to select or engineer the trait successfully, given the secondary effects that would undoubtedly result from differences in so many genes.

Science is only beginning to decipher the relationships between our genes and our physiology and behavior. Researchers have identified a gene or two with connections to a number of complex behavioral traits but know little about the mechanisms or how many genes are involved. With the sequence of the human genome completed, however, a bounty of new information is on the way.

Detailed information about the genetics of religiosity, criminality, intelligence, addiction, and sexual orientation may not come from large gene-association studies specifically designed to explore these connections. Studies in such politically charged areas may prove too controversial to fund. Early explorations of the genetics of human personality and temperament more likely will come from reexamining information from well-funded gene-association studies designed to locate disease genes or to enhance the effectiveness of medications. When these studies have assembled large genetic databases that ensure individual privacy, mining them for correlations with measures of personality and temperament will be possible, though this will require enormous computational power, since many different genes play a role in these traits.

Our task with these early studies will be to distinguish between meaningful associations and mere artifacts of the genetic peculiarities of particular populations. Many of today's results may not withstand the test of time. They analyze too few genes, look at too few people, look for effects from single genes rather than combinations of genes, and are too sensitive to subtle imperfections in methodology. For example, the announcement by Dean Hamer, of the National Cancer Institute, of a "gay" gene at the tip of the X chromosome made headlines in 1993, but his findings were soon challenged by a larger study that did not confirm the homosexual link. Then the subsequent study was criticized. These are wobbly first steps, and at present it would be unwise to make too much of the specific gene linkages that researchers are identifying for complex attributes involving multiple genes. But our knowledge is growing rapidly.

Whether we will be able to devise simple, effective interventions to select complex human traits is uncertain at present, but early hints are promising. With disease predispositions, the feasibility of interventions is clear. Single genes are responsible for many diseases and sometimes contribute strongly even to illnesses with numerous environmental influences. A link exists, for example, between Alzheimer's disease and variants of the gene for a blood protein known as ApoE. People with two copies of a version called ApoE-4 have a 95 percent chance of developing Alzheimer's by age eighty.

Manipulating or selecting personality traits also seems plausible. In 1995, Richard Epstein and his colleagues at Jerusalem's Herzog Memorial Hospital investigated a gene that codes for a cell receptor that binds dopamine, which is a neurotransmitter — a chemical that when released enables nerve impulses to cross the junctions between nerve cells. They found that the length of a region in this receptor gene governs the receptor's ability to bind dopamine and correlates with a person's score on personality tests measuring novelty-seeking. A longer region meant weaker binding by the receptor and a greater tendency for an individual to seek out new experiences and stimulations. This makes sense, because dopamine is the neurotransmitter that is released when someone has sex, eats a great meal, or uses cocaine, and we can feel its pleasurable effects only if it can bind its receptors. Reduce this binding and we might seek stronger sensations, because we need more dopamine to feel the same pleasure. Dean Hamer and his team confirmed Epstein's result and concluded that the gene accounts for perhaps 10 percent of the population's genetic variability in novelty-seeking. Hamer suggested that only a half-dozen genes might be responsible for most of the genetic variability in this trait. But the truth is that today there just isn't enough information to do more than guess about the number of genes that make substantive contributions to any complex trait.

Hamer reported similar findings in other traits. He discovered that the gene for serotonin transporter — a protein that mops up serotonin, another neurotransmitter at nerve junctions in the brain — can predict people's propensity for what is called harm avoidance, a cluster of traits such as anxiety, shyness, and depression. Variation in this gene accounts for less than 10 percent of this genetic predisposition. In the same way that Prozac reduces people's depression by blocking serotonin uptake, these gene variants act like a targeted genetic Prozac to reduce their harm avoidance.

If comprehensive studies show that variants of a relative handful of key genes are responsible for tendencies toward specific personality traits, it may be possible to select or modify embryos to achieve these predispositions. While even a half-dozen genes now seems like a lot to modify, only a few alterations to any particular genetic background might be enough to shift its propensities dramatically because some of the desired alleles would already be present. Selecting an embryo that already has favorable combinations and then refining those through genetic engineering might be especially potent, since nature would do part of the work.

A remarkable experiment in rodents confirms that changing even a single key gene associated with a complex behavior can sometimes be enough to alter that behavior noticeably. By moving a gene from the prairie vole to the mouse, Larry Young and his colleagues at Emory University made the mouse more vole-like in its social interactions. Prairie voles are monogamous. Laboratory mice are promiscuous. Each species has a slightly different version of a receptor for a brain peptide associated with social behavior. When Young inserted the vole gene for this receptor into the mouse genome, the transgenic mice did more grooming and licking in the presence of the brain peptide. In effect, their social behavior moved toward that of the prairie vole. "Although many genes are likely to be involved in the evolution of complex social behaviors such as monogamy," Young concluded, "changes in the expression of a single gene can have an impact on the expression of components of these behaviors, such as affiliation."

Experiments on other complex traits suggest that Young's results are not atypical. In the roundworm, knocking out any one of several genes can greatly increase lifespan. In mice, an extra copy of a particular gene that codes for part of the receptor for glutamate, yet another of the brain's neurotransmitters, enhances learning and memory. In sheep, a mutation of a certain growth-factor gene raises the incidence of multiple births. In humans, a variant of a gene associated with blood-pressure regulation is a virtual prerequisite for the level of endurance needed for high-altitude mountaineering. Individuals with two copies of the right version respond ten times as well to repetitive weight training as those lacking it.

Because disrupting a complex system that is working well is so much easier than improving it, single-gene mutations that enhance the intricate biological orchestration that constitutes us are more intriguing than those which bring disease. When researchers report changing a single gene to enhance memory or endurance or life expectancy, however, we need to be extremely wary. The change may occur only if some particular genetic background is present, or it may have unforeseen downsides. Engineering human embryos in the real world will have to depend on solid animal testing and have to target genes that will not inadvertently affect other traits.

One way of surmounting uncertainties about early interventions may be to duplicate or select for naturally occurring constellations of genes, ones that already exist in the general population and are known to correlate strongly with desired attributes. Such interventions would not require a deep understanding of biological mechanisms, only the ability to select or copy genes.

Because geneticists would be using gene constellations that nature had previously tested in people, risks and unexpected side effects would be small. Essentially, we would merely be creating the technology for transferring intact groups of genes among humans, a trick that bacteria perfected eons ago and now use to spread their hard-won resistances to antibiotics through the bacterial world.

Studying the genes of a single person would help little in identifying such fortuitous constellations. Researchers would have no way to tell which gene variants were relevant to the trait in question. The trick will be to identify the combinations of gene variants common to many people with similar endowments — musical brilliance, athletic excellence, or high energy. Genetic associations that cross boundaries of ethnicity and stand above the noise of natural human variation will be of universal value to us, but nuances that are meaningful primarily to those in specific populations will also surface, the product of more focused genetic studies, like those in Iceland and elsewhere.

Armed with greater knowledge about our genome, we might even be able to capture for our children a hint of the musical talent of a John Lennon, a touch of Einstein's genius, a wisp of the physical prowess of Michael Jordan. At first glance, this notion seems ridiculously simplistic — how could anyone be naive enough to think people will be able to mimic the talents of some great person in their child? But such possibilities may emerge. To do more than guess about their feasibility, we will need to know how many genes are involved in various traits, how effectively we can select and copy them, and how interwoven their effects are.

Many leading scientists are convinced that conscious manipulation of the human germline will eventually be feasible, although they differ about whether it is desirable. "It seems obvious that germline therapy will be much more successful than somatic," said James Watson in 1997. "The biggest ethical problem we have is not using our knowledge." Eric Lander, director of the Whitehead Institute's Center for Genome Research and a key player in the Human Genome Project, agrees about the feasibility, but supports an indefinite ban on such modifications. In 2000, he wrote: "There will come a time when we can do such things safely [germline manipulations], and it is not too soon to ask whether we should."

In the years ahead, parents will face ever more difficult decisions about the genetic makeup of their children, and as with the issues of human infertility, abortion, and assisted suicide, their choices will embody deep dilemmas. Already we are encountering some of these. Using preimplantation screening, we can avoid having a child with certain genetic diseases. Using sperm-sorting technology, we can choose the gender of the child we conceive. And using fetal testing and abortion, we have a brute-force veto over any genetic constitution we choose to avoid. We could outlaw any of these practices, but given the huge deficit of baby girls in India as a result of the illegal use of plain old ultrasound for sex selection, I suspect that such laws would not accomplish much.

When discussing germline manipulation, I also have often mentioned embryo selection. This is because early technologies for directly altering our children's genes will so overlap with coming versions of today's techniques for screening and selecting embryos. Many labels are being used to describe various advanced reproductive procedures: germline engineering, preimplantation genetic diagnosis (PGD), embryo selection, reproductive cloning, germline therapy, germline enhancement. But the distinctions among them often blur when one looks at the outcome of a procedure or at additional technologies that would be used in conjunction with it.

Germline interventions will emerge from, compete with, and depend on simpler techniques such as PGD. Using an artificial chromosome to retard aging might be called either enhancement or therapy and would probably rely on both genetic diagnosis and embryo selection. The profusion of terms and distinctions obscures the straightforward shift now under way. Pure and simple, we are poised to make conscious, highly specific choices about the genetic constitutions of our children and to inject our preferences into the next generation using methods far beyond those previously available.

I refer to this whole realm, which extends all the way from rudimentary embryo diagnostics to germline enhancement, as germinal choice technology, or GCT. "Germinal" emphasizes that GCT manipulates one or a very few of our germinal cells rather than a fetus and is directed toward creating (or germinating) life rather than terminating it. "Choice" acknowledges that our personal preferences will help determine our children's genes. "Technology" recognizes the entry of laboratory machinery into human reproduction to externalize the process of conception.

Choosing Genes

Some parents insist that their children study hard and earn good grades. Some push their kids toward sports. Some want outgoing and popular offspring. Whether we guide our children with a heavy hand or are subtle and indirect, the paths we try to choose for them often tell more about us than about them or who they will become.

The key to figuring out the consequences of giving parents the power to pick their children's genes lies in deciphering not how such choices will affect the children, but how parents will perceive and evaluate these choices. We don't yet know which aspects of personality and physicality can be shaped through GCT, but at this point, any attribute that is significantly heritable is a potential target — and as we have seen, that covers plenty of territory.

Even for highly heritable traits, however, it will be uncertain what a child's unique amalgam of potential and experience will bring. A vision of parents sitting before a catalog and picking out the personality of their future "designer child" is false. One day we may choose our children's tendencies, but we will not be able to choose their actual personalities. Twin studies have shown the importance of our genes, but they've also shown the influences of our life experiences and the role of chance.

Parents will have many ways of influencing their children by guiding their behavior and channeling their development, but the interactions of genes, experience, and environment will not be reduced to simple formulas. However important genetic manipulation proves to be in shaping our children, more familiar tools such as education, mentoring, religion, parenting, and pharmaceuticals will continue to play their roles.

To understand the choices parents may make with GCT, we need to consider some of the dilemmas they will face in deciding whether to avoid temperaments that embody vulnerabilities. Peter

Whybrow, the director of the Neuropsychiatric Institute at UCLA's School of Medicine, points out the complexity in his eloquent book *A Mood Apart:*

> A coding variation that predisposes some families to manic depression — a regulatory enzyme, for example, that promotes an unusually broad oscillation of emotional homeostasis — may in combination with other genes determine melancholic psychosis in one family member, and high energy and short sleep in another, resulting in an optimistic temperament. While both are genetically equally deviant, the phenotype of psychosis marks the first individual as a psychiatric patient and the second as a successful leader . . . On the other side of the coin, the withdrawal and conservation of depression (but *not* the psychosis of melancholy) offers time for reflection, rebuilding, and creative thought. Thus, within the spectrum of mood disorder and emotional temperament may be found behaviors of both positive and negative value to the individual, to society, and in the evolutionary struggle.

The multitude of interactions of experience and genetics that shape a person's nature might create a high risk for severe mania and depression, but individuals with more muted expression or better luck nonetheless may manage to channel their tendencies in extraordinary ways.

None of us wishes to see our children suffer, but if we could protect them from all the dangers and pains of life, we might ultimately diminish them by leaving them untested and shallow. After all, some of the influences that we, in hindsight, most value in our own lives are the painful failures and defeats we have overcome, the jarring potholes that have sent us lurching down new roads.

This sentiment sounds nice, but if we had the power to protect our future child, we might be very reluctant to leave him or her with a predisposition for recurring bouts of dark depression. Nor would the knowledge that our child might use these distressing periods to good purpose make our decision to forgo germinal interventions any easier. I suspect that most parents would make the safe choices and avoid the ragged uncertainties at the edges of human possibility.

The Road to Enhancement

So far, I have avoided the factor most policymakers see as crucial to the future of human germline manipulation and other related technologies — government regulation. I have done so because actions in this area are unlikely to alter the fundamental possibilities now emerging. The legal status of various procedures in various places may hasten or retard their arrival but will have little enduring impact, because, as already noted, the genomic and reproductive technologies at the heart of GCT will arise from mainstream biomedical research that will proceed regardless. Bans will determine not whether but where the technologies will be available, who profits from them, who shapes their development, and which parents have early access to them. Laws will decide whether the technologies will be developed in closely scrutinized clinical trials in the United States, in government labs in China, or in clandestine facilities in the Caribbean.

In discussions of human germline enhancement, the tendency has been to paint in broad strokes. When Lee Silver, a molecular biologist at Princeton University, considers the possibilities toward the end of his thought-provoking book *Remaking Eden,* he suggests that one day we will enhance human cognitive functioning. He also imagines that eventually we may expand our senses by replicating special adaptations found in other animals: ultraviolet and infrared vision (from spiders and snakes), the detection of magnetism (from birds), sonar (from bats), acute smell (from dogs). But Ruth Hubbard, a Harvard professor on the board of the Council for Responsible Genetics, sees no value in such technology. "I would oppose germline interventions even if it were possible to show they are safe," she writes. "As for the notion that we need germline interventions to 'enhance' the abilities we can expect to pass on to our children, I believe that people who cannot deal with the uncertainties implicit in having a child even before that child is gestated are in for trouble."

Many prospective parents may be both intrigued by the possibilities of germline interventions and cautious of them. Taking risks to treat disease is one thing; taking risks to modify an embryo on the way to becoming a healthy child is quite another. Early GCT procedures, such as simple embryo selection, are already occurring, of course, and although limited in scope, they are both legal and safe. As their reach and power grow, those procedures will spread. Eventually, caution may lead parents *toward* embryo selection and germline manipulation, not away from it, because they may worry about placing their children at a disadvantage. Whether parents wish merely to avoid diseases or are reaching for greater stamina, intelligence, and other qualities, they are likely to be moderate in their choices. We have seen just this in the case of human growth hormone in children. The primary users are short children who want to achieve average height. Occasionally, youngsters of average height may take the hormone to try to gain a few extra inches, but irresponsible experimentation has not been a serious problem. Parents do not push bootleg hormone on tall kids, hoping to turn them into goliaths.

In considering how parents may use germinal choice technology as it improves, we must remember that genetic interventions are only a means to an end. If better ways of achieving a result are available, we will use them. The heritabilities of some forms of nearsightedness and farsightedness are around 85 percent, so genetics plays a strong role. But glasses and contact lenses correct these problems so easily that they will not make compelling early targets for embryo selection, much less for germline engineering.

Personal circumstances will greatly influence our attitudes toward genetic interventions. A woman with AIDS would likely find a genetic preventive for the HIV virus an attractive way to protect her child-to-be, while prospective parents without the disease might be uninterested, figuring that AIDS vaccination will be available by the time their child grows up.

Culture and politics too will influence our choices. If you wanted to build a superior human, you would probably choose black skin, at least if the person was going to spend much time in the sun. A comparison of an elderly black person's skin with the wrinkled, damaged hide of an elderly white or Asian sun worshiper shows this clearly. Moreover, the genetic controls of melanin production are probably not complex, since a mere three to five genes seem to determine skin color. But don't hold your breath waiting for such engineering. A genetic module for black skin is unlikely to be in big demand anytime soon. Blacks certainly don't need it, and few nonblacks will want it. Race and parentage carry so much cultural baggage that few parents would use their children as political billboards when sunscreens, clothing, or even drugs and lotions that stimulate melanin production to create natural-looking tans, can protect their skin. People can easily manage with these, and use them to the immediate dictates of fashion, medicine, climate, and whim.

Every germinal intervention will have a different calculus of risks, rewards, and substitutes for parents to weigh. Current targets such as cystic fibrosis, Huntington's disease, muscular dystrophy, and hundreds of other single-gene disorders are relatively uncontroversial because they are obviously therapeutic and require only embryo selection. Interventions to avoid disorders governed by multiple genes lie just ahead. Laboratories now can test a single embryonic cell for variants of a half-dozen different genes simultaneously, so as soon as genomic researchers can identify problematic combinations of genes, germinal screening for them will not be far off.

As more comprehensive single-cell diagnostics emerge, the distinction between diseases and traits that are simply unwanted may well fade. The diversity of our values, philosophies, cultural histories, beliefs, and circumstances suggests that this will be a highly controversial development. People's opinions will differ sharply about what diseases are severe enough to avoid and which extremities of temperament they will reject. And adults who already have these conditions will surely feel strongly about such judgments.

So GCT will have revolutionary implications even before its most potent form, direct germline manipulation, becomes available. Simple embryo screening does much, and questions about its safety are minimal, but its limitations will become evident as parents move beyond efforts to avoid embryos with unwanted attributes and start searching for fortuitous genetic combinations that have predispositions toward "desirable" qualities. As parents try to influence more traits, and those traits involve more genes, the likelihood of finding embryos with favored combinations will fall. Given four possible variants for each of four genes, for example, the chance of some pre-chosen combination arising in an embryo would be only about 1 in 250.

Screening gets unwieldy with large numbers of embryos, because teasing out the single cells needed for preimplantation genetic diagnosis takes a technician about twenty minutes. A way of bypassing this bottleneck will be germline engineering techniques in which specific gene variants are inserted in artificial chromosomes. A way of expanding the capabilities of embryo screening would be to use robotics, of course, but screening would still face a fundamental problem: it can draw solely from parental genes. If a couple do not possess the particular gene variants that would tend to manifest desired attributes, embryo screening cannot help them. The only way parents could give these predispositions to their children would be with germline procedures that pull the needed genes from elsewhere.

Such germline interventions may be therapeutic in that they attempt to enhance a child's health, longevity, or psychological well-being, but rarely will they be a therapy for some specific disease. Embryo selection offers too easy a way of avoiding most potential targets. In essence, germline modification will mean human enhancement, although not necessarily in the sense of capacities pushed beyond what is possible for *Homo sapiens.* Individual parents don't care about the species, they care about their child. For prospective parents, enhancement only means giving their child talents and abilities that he or she could not otherwise expect to achieve. Such enhancement falls within the range of normal human performance and will be much easier technically.

Genes, Dreams, and Memes

In the spring of 2000, I took part in a public forum in Munich, Der neue Mensch (The New Man), about the coming advances in cloning and embryo selection. The brochure for the event, featuring repeated images of a beautiful blonde, nicely captured the specters that haunted the discussion that followed, a dialogue typical of those I have had in Germany, filled with angst over the dangers ahead and burdened by history. This evocation by Germans of their Nazi past brings up the question of whether alluring images in the media of blond-haired, blue-eyed beauties and visions of supermen and perfect humans will drive people's choices about the genetics of their children.

To figure out which traits we will want for our children once we have the power to make such choices, we must think long and hard about who we are. Our evolutionary past speaks to us through our biology and fashions our underlying desires and drives. Our urges are those that best enabled our ancestors to produce as many children as possible and ensure that those children go on to do the same. These acts are the essence of biological success and the heritage of each of us. Simplistic interpretations of what evolution means about who we are and how we should live are frequently misleading and politically self-serving, but when human preferences are relatively consistent across cultures and lands, they are likely tied to our biology, whatever their local accents and nuances.

In order to understand some of the underlying forces operating in our culture, it's useful to state a few generalizations about men and women. (This is risky territory, of course, and we must always be mindful of the many exceptions to any sweeping statements about our natures.) We humans focus much of our energy on sex, on our families and homes, and on our status relative to others. Both men and women are largely monogamous, but not entirely. We favor those more related to us over those less related. We are social creatures and are most comfortable within small groups, of a few people up to a few dozen. We are highly competitive. Men tend to be more aggressive and violent than women. Men compete more for status, power, and wealth — key ingredients in attracting women — and are drawn to beautiful women, which in almost all cultures means women having attributes that signal youthfulness and health, just the qualities needed to bear and rear children. Women tend to care more about how they look — a key ingredient in attracting men — are more socially adept, care more about commitment, and are drawn to men possessing status, power, and wealth — just the qualities needed to provide for children.

The importance we attach to matters such as sex, beauty, status, power, and the success of our children comes from deep within us. We do not have to act upon these urges, but we cannot escape them. It is no fluke that advertisements so frequently employ these themes. They work. Automobile ads aimed at men stress status, power, money, and attracting women; so do ads for computers, vacations, cigarettes, beer, and just about everything else. Fashion ads for women focus on beauty, appearance, sex, and attracting and holding on to the right men.

Many parents-to-be, if they could use germinal choice technology to shape their children, would seek the very attributes that advertisers are pushing. Not everyone would follow stereotypes of beauty, strength, and intelligence, but many would. And we might be surprised at how much we ourselves would be drawn in these directions if given a smorgasbord of qualities to choose from.

Physical attributes will be an enticing target for potential adjustment because people, particularly women, are often judged by their appearance. Cosmetic interventions might involve clusters of alleles that, in the case of women, soften features and bring more symmetry to them, and in the case of men, add to stature and strength. Eventually, self-modification of this sort might lead to somewhat less physical diversity, but we must remember that quite a range exists in what we find attractive. In addition, superficial features such as hair type, eye color, nose shape, and other aspects of our appearance will be easier to modify with hair dyes, colored contact lenses, cosmetic surgery, and clothing than by genetic engineering. Parents are unlikely to consider these lesser attributes seriously except as add-ons to larger packages of choices.

Today, when parents buy donor eggs or sperm, the most important quality they look for in a donor — besides good health, of course — is that ethnicity and basic physical features match their own. They want to be sure that the child appears to be theirs. Beyond those limitations, couples seek the same underlying qualities they might wish for in a mate. Indeed, the catalog listings in sperm and egg donor banks resemble singles ads in newspapers. Here is an ad for sperm: "He has wavy dark brown hair and eyes. He's 5′10″ tall and weighs 156 lbs on a medium frame . . . His ancestry is Eastern European . . . He is currently studying law with a GPA of 3.4. Other interests include music . . . ice-skating, and juggling."

Once we have a band of intelligent devices as assistants, we may not need to be as smart to get along in the world, but we will probably continue to value intelligence. And well we should, since it will no doubt remain a critical asset in our social interactions. It enables us to entertain each other, to play social games well, to compete more successfully with one another. This suggests that in the future we may need to possess a kind of savvy rather than the specialized cognitive abilities or general intelligence that IQ tests try to measure. These two qualities are not the same, but they are not independent either. However imperfect the measures of genetic predisposition for intelligence turn out to be, some prospective parents will want to use them in selecting embryos. Grade point averages and SAT scores are important pieces of information in the biographical profiles of egg and sperm donors in the United States, because they serve as a surrogate for these genetic markers.

Sociability will be another target, because it is so important to positioning ourselves within groups. But for choices of temperament and personality, I suspect that individual parental tastes, in all their diversity, will carry the day. Little agreement exists on whether it is better to be introverted or extroverted, sociable or aloof, cautious or bold, because it depends on the circumstances. Even without concerns about safety, parents will tend to want to moderate temperament and personality. Not only will they often compromise about their own differences in temperament and perception, but they may worry about raising a difficult, inflexible child or wonder if extreme traits will handicap their kids.

This tendency to moderate extremes of temperament may one day render the world a bit less interesting, but germinal choice technology would have to become incredibly widespread to have so far reaching an impact on the human population. In any event, I suspect that we need not worry about being bored by each other in the decades ahead. Moreover, efforts to moderate personality extremes will not rest on genetic manipulations alone. Psychopharmacology too will carry us in this direction, with its antidepressants, mood stabilizers, sedatives, and antianxiety drugs. Dozens of antidepressants, for instance, are now available.

A trend that competes with this move toward moderation will arise as well. People will be inclined to give their children those skills and traits that align with their own temperaments and lifestyles. An optimist may feel so good about his optimism and energy that he wants more of it for his child. A concert pianist may see music as so integral to life that she wants to give her daughter greater talent than her own. A devout individual may want his child to be even more religious and resistant to temptation. To the extent that enhancements of this sort by parents engender mindsets disinclined to attenuate the traits in their own children, such traits may reinforce themselves from generation to generation and push the limits of genetic possibility and technical know-how. These countervailing tendencies, toward restraint on the one hand and intensification on the other, could produce a relatively unaltered core population and clusters of individuals with highly specialized talents and temperaments.

Our emotions play a large role in regulating our behavior, so we need to consider their nature in thinking about our future choices with germinal technologies. Biology uses emotion to channel behavior in ways that have been evolutionarily fruitful. We do things because they gratify our desires, calm our fears, stop our pain, feed our hunger, satisfy our curiosity, make us feel happy. Generally, if we seek out things that give us pleasure and avoid situations that cause us pain, we have a good shot at feeding ourselves, having kids, taking care of them, and being the best biological creatures we can be, which means leaving the most kids behind. We love sex, so we devote time and energy to making sure that it happens. We fall in love and want to have a family. Pretty soon, kids are running all over the place. "Great!" says evolution. A man gets aroused, has a little fling, and leaves an extra kid somewhere. "Cool," says evolution. A wife gets lonely, seduces some famous rich fellow, has his baby, and her husband is completely oblivious. "Not bad," says evolution; the kid may have picked up a useful gene or two. Leaving our family wracks us with grief, so we stay with them. Letting someone kill our child would tear us apart, so we do everything we can to keep that from happening. We yearn to feel important, so we work hard and plot to get ahead. We love our kids and want them to do well, so we help them.

Our biological urges have guided us well in the past, but they are slow to adjust to change. These days, contraceptives are common, so we have sex and don't have children. We like sweets, but sugar is everywhere instead of just in fruit and honey, so we get fat. We want to be happy, so we drink alcohol and end up with a wrecked home. Smoking brings us a rush, so we end up with cancer or heart disease. The pitfalls are many, and they will only multiply. As we learn to stimulate our pleasure centers with drugs that satisfy our urges chemically and uncouple our emotions from our behavior, questions will arise: who will we be, and what will motivate our lives? Our future may be as disorienting as it is fulfilling.

Some critics of germinal choice technology have painted nightmarish pictures of totalitarian governments breeding armies of obedient soldiers, but the collective dynamics of free individual choice are likely to channel us far more effectively than any dictator. As we start to transform our biology, extending our lifespan, altering our personalities, and changing the ways we gratify ourselves, it is not at all clear what we will seek and where the commonalities among humans will lie. Certainly, the individual honed by genetic engineering to religiosity and introspection will want something far different from the one with heightened aggression and competitive urges, or the one with extraordinary beauty and grace, or the one with heightened sexuality and carnal appetites.

Memetics — the study of the cultural elements, or memes, that propagate among us — offers a framework for assessing which genetic constructs are likely to proliferate in the human population. Our preferences in the gene modules we place on artificial chromosomes are memes that will spread horizontally through the adult population as well as vertically from parent to child. We are the vessels that will carry these genes, but their spread will depend not on our reproductive success, but on how well they can insinuate themselves into our hearts and minds to gain new users. These modules will be in a game of their own, competing with each other for our attention. Those that succeed — by aligning well with human desires and by securing the right publicity — could spread rapidly.

This competitive process seems more technological than biological. Devices gradually improve, different versions compete, and occasionally discontinuities crop up as new approaches render old ones obsolete. The similarity should not surprise us, because we are dealing with technology here, albeit biological technology. In essence, genetic modules will be recreating the rapid memetic expansions that have characterized successful consumer products.

Future human enhancement will favor those traits and predispositions that people are most likely to copy and adopt as their own. From one perspective, genetic constructs will eventually be no different from other products promoted by advertising and influenced by word of mouth. From another, however, gene modules are unique, because they will alter the biological natures of children and affect their future propensity to apply the technology to their own children.

Broad screening of embryos to avoid diseases and other unwanted traits will probably spread quickly once it becomes readily available. Similarly, we are almost certain to embrace the germinal enhancements of added health and longevity when they begin to emerge. Our yearning for immortality has been too persistent and potent for us to do otherwise. But assessing which modifications to a child's temperament will appeal to parents, or which enhancements of a talent such as music, art, or athletic ability they might choose, is mere guesswork at this point.

These preferences too will spread mimetically, although most couples drawn to germinal choice may pick and choose among their own temperaments and personalities, enhancing qualities they like in themselves and moderating ones they don't. These decisions remind me of a comment attributed to George Bernard Shaw. Supposedly, when a beautiful opera singer suggested to him that they have a child together, because his brains and her looks would be a wonderful combination, he responded, "But what if it had my looks and your brains?" In the future, prospective parents may try to avoid such unintended results, instead striving to pass on what they feel is best in themselves and their mates. And with the decline of two-parent families, single women using select sperm from donors may be making these genetic choices.

We should be suspicious of simplistic interpretations of the impact of germinal choice. Our reactions to it will surely be as complicated and unpredictable as the rest of life, and equally driven by individual temperament, personality, and circumstance. The only real way to gauge our distant choices may be to observe our immediate ones as GCT expands, but it will not be easy to extrapolate from these early choices, because our individual expectations and the demands of society may shift substantially as the technology evolves.

One thing we can count on, though, is that any combination of personality and temperament that predisposes people to embrace biological selection and enhancement will be highly represented among those who use germinal choice. To the extent that the personality attributes that lead to this are genetic in nature, the technology is likely to reinforce them in successive generations. Enhanced humans will manifest and reinforce their philosophy in their biology.

7

Ethics and Ideology

> It was the best of times, it was the worst of times, it was the age of wisdom, it was the age of foolishness, it was the epoch of belief, it was the epoch of incredulity, it was the season of Light, it was the season of Darkness, it was the spring of hope, it was the winter of despair, we had everything before us, we had nothing before us, we were all going direct to heaven, we were all going direct the other way.
>
> — Charles Dickens, *A Tale of Two Cities*

ONE DAY we will manipulate the genes of our children in sophisticated ways using advanced germinal choice technologies. In spite of a general uneasiness about such technologies, we will likely use — and misuse — them as soon as they arrive, just as we have earlier breakthroughs along this path. Prospective parents use ultrasound to determine the gender of a fetus so they can abort a baby girl, amniocentesis to look for the telltale extra chromosome 21 of Down syndrome, and preimplantation genetic diagnosis to select an embryo free of cystic fibrosis. Single-cell diagnostics are still too rudimentary and expensive to make broad PGD screenings feasible, but the desire and the perceived need are clear.

The dilemmas surrounding the technology are not completely new. In 1985, Lee Ann Currie and her husband learned that their newborn first child, Natalie, had Fanconi anemia, an inherited disorder that usually brings cancer and death long before adulthood. Natalie's best hope was a transplant of cells from a genetically matched donor, so Lee Ann decided to conceive more children in the hope that one might have the right genes. Four years, three pregnancies, and two daughters later, she succeeded with the birth of Emily. Natalie had a new lease on life. The Curries' situation was no abstraction; they were balancing risks and rewards to try to save their daughter. Since the transplant cells would come from blood in the umbilical cord, which posed no risk, the calculus seemed easy to them. Criticism of her motives brought no apologies from Lee Ann: "We live in America, and we can conceive a child for any reason that we choose," she said. "How dare anybody judge us for wanting to conceive a child to help save the life of another one."

Fast-forward fifteen years, to August 2000. Linda and Jack Nash's six-year-old daughter, Molly, suffered from the same deadly anemia, and the Nashes, like the Curries, were hoping to bear a healthy child who could donate umbilical cord blood to Molly. When, after five failed IVF attempts, Linda finally delivered a healthy son, success was theirs. That the Nashes had used PGD to screen their embryos and reject those with the anemia wasn't newsworthy. Many couples had done as much. What attracted international attention was that they took the further step of selecting among healthy embryos to be sure that the cord blood would be a match for Molly. By going beyond avoiding disease, they had pushed into the frontier of genetic testing. But the Nashes' goal was the same as the Curries', and so was the result.

As couples with less agonizing problems begin to grapple with the decision to use PGD and other advanced reproductive technologies, such tests will cease being philosophical abstractions and become mere medical tools, the means to diverse and very personal ends. And in the struggle to reconcile our values and philosophies with personal necessity, consistency often goes out the window. Senator Strom Thurmond of South Carolina, a vocal opponent of abortion, supported embryo research years before other conservative pro-lifers began to mute their opposition. The plight of his own daughter, who has juvenile diabetes, brought home the toll that research restrictions can exact if they delay the discovery of life-saving medical interventions.

Such incongruous views are not unusual. In July 1991, fifty-two-year-old Arlette Schweitzer gave birth to two healthy babies, Chad and Chelsea, and became the first American grandmother to give birth to her own grandchildren. The babies were to be raised by Arlette's daughter, who was born without a uterus. You might assume that Arlette would strongly support surrogate motherhood, but she does not. As an ardent Catholic, she is deeply troubled by the idea of a woman being paid to carry another's child and by the use of donor sperm or eggs. To her, there's something wrong with "borrowing the actual makings of the baby from other people." She sees what she did for her daughter as different: "I don't even consider myself a surrogate," she says. "I was just a mother helping her daughter and would never have considered doing it under other circumstances."

Interpreting rigid dogma so as to allow new reproductive technology is quite an art. The Catholic Church's categorical opposition to masturbation, for example, presents a problem for the devout who are seeking a child through *in vitro* fertilization: how can a man collect the needed sperm without sinning? Despite official opposition to IVF, some in the Vatican say a man may obtain sperm by using a condom perforated by a pinprick, so that conception is still possible and intercourse remains a "proper marital act."

Contortions of this sort will be commonplace once full-blown embryo selection and germline technology are available. Preimplantation screening will not resolve our basic dilemmas; in fact, screening will deepen and multiply them, transforming them into questions about embryos rather than fetuses.

Some nations may hold off the technology for a while, but the long-term impact of such bans will be no greater than previous such efforts. Germany, haunted by its Nazi past, opposed genetic technology for many years and in 1991 enacted an Embryo Protection Law that was the most restrictive in the world. By 1993, however, the realization had sunk in that biotechnology would pass the country by, and Germany moderated its tight restrictions on great swaths of genetic research. In 2000, the nation went further and began to debate the watering down of the 1991 law. Switzerland, home to many pharmaceutical companies, seemed especially unlikely to lead a charge against genetic medicine, but in 1997 it almost did. The Swiss nearly passed a plebiscite to ban such research. They blinked only when the economic costs of driving drug companies abroad became clear, and in 2000, another plebiscite there, this time to ban IVF, lost by a margin of more than 30 percent.

These economic forces operate throughout the developed world. That we will halt the global scientific effort to elucidate human genetics is inconceivable. And we will have no trouble figuring out how to justify using more potent germinal choice technologies as they emerge. Bans in this or that country surely won't keep them from spreading. When large numbers of people want something that regulators cannot monitor and that small laboratories in any country can provide, people obtain it.

Though the prospect of genetic manipulation disturbs William Gardner, a bioethicist at the University of Pittsburgh, he argues convincingly that no ban can stop it: "Both nations and parents have strong incentives to defect from a ban on human genetic enhancement, because enhancements would help them in competitions with other parents and nations. The ban on enhancement, moreover, is vulnerable to even small defections because the disadvantages of defecting late will increase the incentives for non-defectors to follow suit, causing defections to cascade."

Couple Gardner's argument with the global diversity of attitudes about germline selection and enhancement, and the writing is on the wall. The inevitability of these technologies, however, doesn't mean that no one will try to block them or that no reasons exist for concern about their possible risks. Although many people believe that the potential benefits of these technologies outweigh their dangers, many do not share this belief. Some assert that we don't have the wisdom to shape our children in this way, others feel that we are wrong to play God. Some fear that the technology might lead to genetic discrimination, be abused by tyrants, or enable the wealthy to give their children superior talent and leave the rest of us behind. Still others worry that it might corrupt the relationship between parent and child, transform children into mere objects, or burden them with unrealizable parental expectations. In short, many people are uncomfortable with the idea that we might take control of our own evolution, and many of them would like to stop this technology.

Leon Kass, who was appointed by George W. Bush in 2001 to chair the President's Council on Bioethics, is vehement about the dangers he sees. In an article in the *New Republic* in 2001, he draws a line in the sand, opposing the cloning of human embryos for stem cell research in fear that it might lead to reproductive cloning. "We are compelled to decide nothing less than whether human procreation is going to remain human," he writes, "whether children are going to be made to order rather than begotten, and whether we wish to say yes in principle to the road that leads to the dehumanized hell of *Brave New World*."

Public policy in Germany and France takes a similar slant, banning human germline engineering categorically, labeling it at various times as an assault to "human dignity" and a violation of our "right" to an unaltered genetic heritage, and saying that it would modify the human gene pool, the patrimony of all humanity. In 1997, the Council of Europe wrote a "Convention on Human Rights and Biomedicine," asserting that "an intervention seeking to modify the human genome may only be undertaken for preventive, diagnostic, or therapeutic purposes and only if its aim is not to introduce any modification into the genome of any descendants." The United Nations Educational, Scientific, and Cultural Organization was less adamant in its recent Universal Declaration on the Human Genome. UNESCO did not forbid germline intervention, but in deference to German concerns, called for further study of practices that could be "contrary to human dignity."

The debate between those who would block such reproductive technology and those who would encourage or at least tolerate it brings up our most fundamental beliefs about who we are and what we value. Glenn McGee, a respected bioethicist at the University of Pennsylvania, thinks that public education and discussion will lead to consensus on cloning and germline manipulation.

"Cloning must be banned at least in the short term," he said in one public forum, "not because it's bad in principle, or sinful, or dangerous, but because there is no collective wisdom yet in our common conversations about cloning and reproductive genetics, no consensus about what liberties people can take in altering their offspring." The hope of attaining consensus in this area seems highly unrealistic. Even ignoring safety issues, consensus on cloning, much less genetic enhancement, will be harder to achieve than agreement on abortion, which seems far from our grasp.

We would do well to explore the arguments for and against advanced reproductive technologies not with the extravagant hope of resolving our differences, but with the modest one of clarifying them. We will do well if we can figure out how to come to grips with these differences and balance the opportunities and dangers the technologies embody. The importance of our efforts does not lie in whether we decide to allow such technologies; they will arrive anyway. The point is how much they will rend our society in the process.

I have grouped the diverse arguments about the emerging technologies into intuitive categories, separating moral objections, for example, from technical ones. Not surprisingly, we will have to approach our disputes in these different realms quite differently. We typically find moral and religious compromise more difficult than compromise about technical oversight. Sometimes, in our unwillingness to acknowledge fundamental disagreements over values, we infuse these beliefs into other areas, disguising them.

The assertion of a human right to an unaltered genetic constitution, for instance, sounds scientific and political, but where does this "right" come from? The assertion is spiritual, and virtually identical to the declaration that we should not play God. One cannot rebut this as a religious belief, but it is unconvincing in secular garb. The normal workings of natural selection can be harsh, and the blanket prohibition against gene therapy is too at odds with accepted medical practice. It seems inconsistent to oppose embryo gene therapy that would protect against the degeneration and early death of children with Tay-Sachs yet to have no qualms about heart surgery to save a newborn. Genetic diseases can be every bit as brutal as nongenetic ones.

Most of us would shake our heads in disbelief at anyone who argued that a child had an inherent right to an unaltered biological constitution and should undergo no surgical procedures before adulthood. When we hear the same argument against genetic manipulation, however, we take it seriously, though it is just as out of line with our values. After all, when fetal testing reveals cystic fibrosis, some 90 percent of couples in the United States choose to abort, and even more would probably avoid implanting a seriously afflicted embryo.

The Perils of Germinal Choice

The most difficult type of germinal choice technology for people to accept is germline enhancement — the direct manipulation of an embryo's genome to improve it in some way. Although as embryo screening becomes more sophisticated it will be able to match many of the immediate possibilities of direct enhancement, more complicated enhancements no doubt will require direct interventions, so in this discussion I will focus mainly on these when speaking about germline enhancement. As the ultimate example of GCT, it will no doubt figure in any serious reworking of human biology, and except for cloning, it has been the most criticized future reproductive technology.

Arguments against human germline enhancement rest largely on assertions that it is morally wrong, that it is too dangerous, that it will be badly abused, or that it could bring dire indirect consequences in personal, political, social, environmental, or spiritual ways. Let's look at each of these.

The most common form of the first assertion, that germline enhancement is morally wrong, is: we should not play God. But there are many secular variations. Along with our "right to unaltered genes" are the ideas that children should not be manufactured, that our gene pool is the common property of all humanity, and that genetic manipulation would assault human dignity.

As for playing God, by the measure of earlier ages, we do just that every time we give our children penicillin, use birth control, fly in an airplane, or telephone a friend. We embrace technologies that tame and harness nature because we think they improve our lives, and we will accept or reject human genetic manipulation on the same grounds.

But perhaps the scale of primitive humans is no longer appropriate, and playing God is a rather hyperbolic way of speaking about any of our tinkering with the natural world. At a 2001 conference on cloning, Rabbi Moses Tendler, a professor of Jewish medical ethics at Yeshiva University, spoke against reproductive cloning, but not because he considered it playing God:

> God gave us molecules. God gave us atoms. We put them together differently. We are not playing God by doing that. We can't get along without Him . . . God is the source of all science and God is the source of religion, and God is not schizophrenic. He doesn't fight with Himself. If there is seemingly conflict between the two, it's based upon one of three possibilities. We don't understand what God said, we don't understand the science, or, the usual explanation, we don't understand either.

Metaphors about "human manufacturing" are another way of articulating that germline manipulation would somehow violate the natural order. The infusion of conscious human choice into the process of conceiving a child blurs the line between the biological and the technological in the same way that artificial intelligence does, but genetic engineering is not about to turn our children into manufactured products. A freely chosen nine-month pregnancy is nothing like the controlled and optimized assembly-line manufacturing evoked by this metaphor. Moreover, children, whatever their genetic makeup, are far too influenced by the vagaries of individual experience to be anything but unique and highly individual.

Communal ownership of the human gene pool is an even stranger concept. The gene pool is a conceptual abstraction that is simply the sum total of all the genes of the reproducing population. We affect the gene pool every time we save a diabetic who would otherwise die before reproducing, every time we bring a child into the world, every time we inoculate a child, protecting him or her from fatal infectious diseases. Do those who argue for collective control of our gene pool imagine that humanity as a whole should oversee these choices as well? Such invocations of the sanctity of our gene pool are not scientific but religious arguments. John Fletcher, who was the first chief of the bioethics program at the National Institutes of Health and is now professor emeritus of biomedical ethics at the University of Virginia, commented in 1998:

> The idea of natural law is one that I think is not a viable concept when it comes to the gene pool . . . Suppose we really knew how to treat cystic fibrosis or some other very burdensome disease and didn't do it because of the belief that people had a right to an untampered genetic patrimony. Then, you met a person twenty-five years later and did the Golden Rule thing and said, "Well, you know, we could have treated you for this, but we wanted to respect your right to your untampered genetic patrimony. Sorry." It doesn't take a highfalutin ethicist to realize that's just plain wrong. You violate one of the basic principles of morality, namely that you want to treat a person as you would want to be treated.

The moral and religious arguments against genetic manipulation are unconvincing to me, but they have significant sway. Beliefs about right and wrong are deep-seated, and I would be astonished if the coming extension of human control into the intimate realm in which life passes from one generation to the next did not provoke strong reactions.

James Watson was instrumental in bringing the Human Genome Project into being. When I asked for his reaction to the idea of the "sanctity" of our genome, he couldn't have been more blunt. "I just can't indicate how silly I think it is," he said. "I mean, sure, we have great respect for the human species. We like each other. We'd like to be better, and we take great pleasure in great achievements by other people. But evolution can be just damn cruel, and to say that we've got a perfect genome and there's some sanctity to it . . . I'd just like to know where that idea comes from. It's utter silliness."

This is not a battle between science and religion. French Anderson, the physician who performed the first human gene therapy, categorically opposes germline enhancement on moral grounds, while some theologians are quite open to it. Rabbi Barry Freundel, of the Georgetown Synagogue in Washington, D.C., and a consultant to the Presidential Commission on Cloning, writes:

> I do not find [human beings gaining control of their own evolution] to be any more troubling than discussing any other human capacity to alter the natural world. I take this position as a result of Judaism's teaching that human beings are the most important part of G-d's created universe . . . G-d has entrusted this world to humankind's hands, and the destiny of this world has always been our responsibility and our challenge. Whether or not we live up to that challenge is our calling and essential mission.

A philosophical debate among theologians, ethicists, and scientists about the morality of embryo selection and manipulation is unlikely to change many minds. When the words die away, people usually find themselves pretty much where they started, especially if the arguments are abstract. But when discussions move from the theoretical to the messy details and contradictions of the real world, greater empathy and acceptance often arise. Not surprisingly, our ideological rigidities soften still more when an issue touches us personally, as was the case with Senator Thurmond and embryo research.

No amount of debate about advanced reproductive technology will settle the question of whether human genetic enhancement is right or wrong. But in pluralistic societies, this situation is not new or even unusual. There are many divisive, value-laden issues concerning sexual orientation, religion, and lifestyle. We have come to rely more and more on tolerance and individual discretion in these areas, and I suspect that eventually we will also have to do the same with advanced reproductive technologies.

The morality of destroying human embryos is an example of particular relevance to GCT. At heart, this is a theological debate about the onset of personhood and will probably never be resolved. Many practicing Catholics now see conception as the beginning of life, but this view dates back only to 1869, when Pope Pius IX removed a three-century-old distinction between an "unensouled" and an "ensouled" fetus and declared that God infuses each individual with a soul immediately after conception, instead of months into the pregnancy. Judaism fixes the beginning of human life at about a month after conception, and Islam somewhere between one and three months. Rather than impose a particular religious view on a diverse population, secular societies such as the United States have developed a body of laws on abortion, birth control, and *in vitro* fertilization that allow the exercise of individual discretion in the fate of embryos and early-term fetuses. But conflict is bound to intensify, as the destruction of embryonic cells becomes more frequent and openly accepted in advanced reproductive procedures and medical research.

The next assertion, that germline enhancement is too difficult and dangerous, is not a valid argument against exploring the technology. The complexity of the human genome is obvious to any geneticist. Six months after completion of its rough draft in June 2000, geneticists still disagreed about the total number of genes. Reports in early 2000 that there were only some 30,000 of them came as a shock to scientists who had thought there would be 100,000 or more. Not only do the influences of most of our genes depend on the nuances of the surrounding genetic background, many genes code for several different proteins. Some observers throw up their hands at this thicket and say that untangling our genome enough to be able to shape the traits of our children is fantasy. Others admit the possibility, but believe that the dangers and uncertainties will always preclude safe germline manipulation in humans.

Dr. Nelson Wivel, who served on NIH's Recombinant DNA Advisory Committee when it approved the first gene-therapy trials, said as much in 1993:

> The risks of the technique will never be eliminated, and mistakes would be irreversible. Germ-line gene modification will always be associated with the risk of unpredictable genetic side effects, and for this

> reason it never should be approved for use in humans. Whatever the mechanisms of review and approval, they are not likely to be fail-safe because it is not possible to guarantee safety and reproducibility in biological systems. Further, there is the ever-present potential for the delayed appearance of unpredicted side effects that could be passed on to future generations.

No knowledgeable person denies the complexity of biological systems, so a dash of skepticism amid the exuberance of daily headlines about the genomics revolution is welcome. But to conclude that we cannot surmount the technical and scientific obstacles is premature, to say the least. At this time, human germline manipulation is not feasible or safe. A decade from now, it still won't be. Two or three decades hence, however, the story may be different. Viable germline interventions in humans will require no fundamental breakthroughs, only a steady increase in the scale of our exploration of the human genome. Within ten years, we will know much more about how our genetic predispositions and vulnerabilities manifest themselves. Many of these influences will probably be impossible to manipulate using current technology, others will prove difficult to decipher but not unmanageable, and still others will be relatively easy to change.

We have barely taken the first step in our coming journey, yet we already know the gene variants we might copy to lower our risk for diseases such as cancer and Alzheimer's. Scientists are more likely to uncover a host of genetic manipulations with meaningful, predictable consequences than to find none. But even if nothing useful is achievable by directly engineering our germlines, we still will have to face the challenges of germinal choice. Safety is not an issue when it comes to germline selection based on preimplantation genetic diagnosis. Indeed, PGD will be in the vanguard of germinal choice, at least for the next couple of decades, and will bring us the same ethical questions as germline engineering.

If germline engineering never becomes safe enough for human use, this will pose little risk. Efforts to modify human embryos might still occasionally take place, but these failures would be a minor problem compared with the many thousands of instances of fetal damage from alcohol and drug abuse and trauma. Our failure in germline research might be a letdown to those who hope that medical science will enable us to transcend our biological limits, but previous generations have come to grips with these, and I daresay we could too. Many people would heave a sigh of relief knowing that science fiction would remain fiction, that human biology would remain the biology we know, and that the present whirlwind of change would not drag loose this anchor of our identity.

In short, if our science proves incapable of reworking human biology more deeply than will occur through embryo selection, no ban on germline engineering is needed. Only if such interventions offer a safe and reliable path toward a substantive enhancement of human traits — the addition of vital years to our lives or improved physical and mental functioning — will we face difficult decisions. Success, not failure, is what will force us to decide when to use this powerful tool and how to deal with its personal, social, and political consequences.

The assertion that human germline engineering can't work is irrelevant either to debate about what to do if it does or to discussion of the similar possibilities that PGD will bring. The argument that germline interventions are too dangerous needs to be rephrased as a question: what level of safety should these interventions attain before they will be a viable extension of preimplantation genetic diagnosis? And to be useful, the discussion needs to be concrete. Grappling with how much risk to tolerate for specific interventions is what will spark the most constructive and vigorous debate.

The third assertion is that even if germline enhancement were beneficial in some ways, the technology would invite too much abuse. Critics who take this tack fear that the government will initiate eugenic programs to shape the genetics of its citizens. They worry that parents, perhaps seduced by advertisements or competitive pressures, will make dangerously mistaken choices for their children. We should not discount such fears, but we must keep in mind that conjuring dark scenarios for any potent new technology is easy.

Computers and telecommunications have brought us great benefits, but they have also opened up new avenues of surveillance and intrusion.

Although the eventual possibilities of direct germline enhancement are more potent, embryo screening is thc germinal technology that will soon offer large numbers of parents the opportunity to make choices about their children's genes. Thus, thinking about how to handle the emerging powers of preimplantation genetic diagnosis today will help prepare us to deal with more challenging choices tomorrow. To decide how much we will allow our fears about potential abuses of GCT to influence our current policies, we must weigh the relative risks and rewards, and gauge whether they justify bans or only vigilance.

In assessing the tradeoffs, we should keep in mind that menacing visions of abuse implicitly assume the technology is potent and effective. Misuse of ineffectual technology is hardly a threat, so if we really do need to block GCT to protect ourselves, we will be forgoing the substantial benefits inherent in its success. We mislead ourselves if we imagine that the tradeoff is between meager benefits and great dangers. The more potent the technology, the more possibilities exist for both use and misuse.

To judge the desirability of trying to forestall any germinal choice technology, we must also determine whether the approach would actually reduce the likelihood of abuse. In the short term, of course, it would. Anything that slows the arrival of a technology delays the day when we have to deal with its consequences. But if advanced germinal technologies will eventually arrive anyway — and even critics generally agree that they will — we must ask which is safer: a path that drives the technologies underground and out of public view, or one that explores them openly; a path that pushes their development into the hands of rogue scientists and states, or one that keeps them in the scientific mainstream; a path with surreptitious funding inspired by visions of black-market profits, or one with aboveboard financing and open-market incentives and constraints.

The answers seem clear. Keeping nascent germinal technology in the open and exploring it in full view lowers rather than raises risks. Openness minimizes surprise developments, warns us of impending social challenges, and allows us to gather information about problems that surface during early implementation. Such cautionary knowledge, which is particularly important with uncertain procedures such as germline manipulation, comes cheaply for society, because while the technology is in its infancy and too expensive and rudimentary to be used much clinically, only small numbers of people are at risk. A critic might argue that even a single damaged child would be too many, but if that well-intentioned idea were to drive policy, we'd have to eliminate cars, ban sports, stop vaccinations, and avoid reproduction altogether.

Given the gruesome experiments conducted in Nazi concentration camps, compulsory sterilization laws in the United States, and the pseudoscience of early eugenics when its philosophy was generally accepted, we are right to worry about tyrannical abuses of germinal choice technologies such as PGD and germline intervention. But tyrants can draw on far simpler technologies to abuse their populations and consolidate their power. Guns, jails, and torture chambers do their work well. Even if we consider only biotechnology, selection and manipulation of human embryos is not the biggest threat of totalitarian regimes. We have much more to fear from the weaponization of smallpox, bubonic plague, or other ancient foes that once infected and killed millions. Advances in medicine and public health had nearly vanquished many of these, but if some nation or terrorist group uses these natural enemies of human life, strengthening and spreading the deadly agents, the consequences could be catastrophic. The anthrax attacks in the aftermath of the World Trade Center disaster of September 2001 have given us a warning.

With reproductive technology, at least we are safe from abuses as long as we preserve our own institutions and freedoms. With freely chosen technologies, our only immediate dangers will be of our own making. The mistakes or malign intentions of others will have little direct impact on us.

We often have great confidence in our own abilities to use technology responsibly, but little faith in the wisdom and motivations of others. Parental abuse of embryo screening, though a reasonable concern, need not be addressed until specific practices and choices that are clearly problematic begin to materialize. Protecting against nebulous future possibilities by walling off entire technologies is a mistaken approach. GCT will not pervert our desire to give our kids advantages; it will simply give us another tool, perhaps a powerful one. Again, most parents are apt to be quite cautious in applying the technology. They will make mistakes, but so would courts and review boards. At least parents know that they will suffer the consequences. If a child is diminished, the family will have to deal with that. The child may suffer most, of course, but the parents will also bear a heavy burden. Every day they will have to care for the child, face their fractured dreams, and live with their guilt. A judge suffers no such consequences. Society as a whole, however, does incur future costs and therefore will almost certainly be inclined to constrain the options of parents who use GCT.

An Eventual Reckoning

The final assertion is that the ultimate consequences of manipulating our genes will be too dire because obscure and distant dangers lurk behind the revolutionary technologies. Without question, the personal, social, political, and philosophical consequences of human control over our biological destiny are enormous, and no one can know where this path will eventually take us. But we must decide whether this implies that we should try to stop this technology or even slow it down.

Procedures with obvious flaws are not the real danger, because we can see and respond to them before the interventions become widespread. A few injuries might occur, of course, but probably far fewer than from other medical errors today. Adverse effects that lie undetected for decades or generations are what most worry bioethicists, because by the time we notice them, many people might be affected. Such concerns about conscious manipulation of our genetics are numerous:

- People might turn into biological time bombs. Perhaps our descendants' bodies will fail at some point, like flawed mutants in a horror movie.
- Our genetic constitutions might become impoverished. Perhaps human biological diversity will diminish and leave us vulnerable to plaguelike diseases. Gene variants whose importance we have not yet grasped may be lost, reducing the creativity and the drive to succeed kindled by human imperfection.
- Society might fragment. Perhaps the rich will buy the best genes for their children, creating a chasm between themselves and the mass of humanity. Perhaps society will discriminate against an underclass of the unenhanced, as envisioned in the 1997 film *Gattaca.*
- Our relationships and values might become distorted. Perhaps parents will view their designer children as objects and burden them with their expectations. Perhaps children will see their lives as diminished and circumscribed by their genetic pedigrees. Perhaps our empathy for those with various illnesses will fade as these afflictions become preventable.
- We might lose our spiritual mooring. Perhaps our mortality and imperfections are what give our lives meaning.

This list, though far from comprehensive, includes most of our inchoate concerns about human genetic design. Underlying them is a great uneasiness that we are opening a door better left closed, that we have neither the wisdom nor the knowledge to handle such power responsibly, that no free lunch is possible, that we will eventually have to pay for our hubris.

The first fears in the bulleted list are essentially medical. As we have seen, worries about safety are nearly nonexistent for simple embryo screening. But germline manipulation is another story. Even after it has been thoroughly tested in animals, the possibility of subtle, unanticipated complications in humans will remain, given the difficulties of testing embryonic or fetal interventions. Any intervention, however, whether genetic or not, can have unanticipated side effects. In 1945, women began to use diethylstilbestrol (DES) to prevent miscarriages, and twenty years later, a rare form of vaginal cancer began to show up in their daughters — young women exposed to DES as fetuses. In 1971, the Food and Drug Administration banned the use of the drug by pregnant women, but by then at least 2 million babies had been exposed. Some 500 cases of the cancer have since appeared.

Every medical treatment has risks. Ironically, the only way to establish safety is through use, and the less obvious the problem, the more widespread the use will have to be to uncover it. Say we find a miracle vaccine for the common cold, and that for every 100,000 people who take it, one drops dead. An initial screening of 1,000 volunteers would have a 99 percent likelihood of showing complete safety, because there is only a 1 percent chance of including a susceptible individual. But were the vaccine rolled out to all grade school kids in the United States, some 300 would die.

One way of dealing with such hazards is to set extremely high hurdles for safety in elective procedures, allowing unproven ones only in cases of dangerous or life-threatening conditions. But people's willingness to accept risk varies, and so does the way people evaluate what is tolerable. Hormone replacement therapy following menopause may seem frivolous to one woman and a necessity to another.

A different approach is to establish likely safety in early tests and then gradually expand usage in carefully monitored studies. We try to do this today, though often with inadequate monitoring after large clinical trials have been completed. Once drugs and procedures pass clinical testing and gain FDA approval, they come into general use. At this stage, rare but serious side effects occasionally show up, forcing a drug to be withdrawn from the market, as occurred with Rizulin, the diabetes drug that sometimes caused fatal liver damage.

In vitro fertilization is an example of a real-world rollout of a procedure in reproductive medicine. The success of Patrick Steptoe and Robert Edwards in 1978 brought enormous publicity, but usage rose gradually. The first IVF birth in the United States did not occur until 1981, and even by 1989, doctors were performing only 8,000 IVF procedures a year, compared with 100,000 now. Slow adoption of the technology was not the intention of the IVF clinics, since they promoted the procedure aggressively; it resulted from reluctance on the part of patients, because of the intervention's high cost, low success rate, and grueling regimen.

A drug can achieve wide use much more quickly, as shown by the public's enthusiastic embrace of contraceptive pills, Viagra, and Prozac. Such rapid rollouts are dangerous, however, because they don't allow time for all potential problems to show up. Use of the diet drug fen-phen, a combination of fenfluramine and Phentermine, soared following glowing reports in 1992 of its use in achieving substantial weight loss. By 1997, when evidence of heart valve problems appeared and the FDA pulled the drug from the market, doctors had already written some 18 million prescriptions.

When use is likely to spread rapidly, stringent testing is critical, but direct germline enhancement, as opposed to enhancement through embryo screening, will probably have a slow rollout. Initially, such interventions will be expensive and couples will worry about the uncertain risks and unproven benefits. Only a few wealthy, highly motivated, and relatively adventurous technophiles will have both the interest and the means to use the procedures. Only when costs fall, risks become clearer, interventions become more potent, and the procedures more familiar will usage take off. This will provide the time to identify subtle, delayed problems before germline engineering spreads.

If we block these early adopters or drive them underground, however, we might well slow initial usage, but the problems that impede early acceptance — technical limitations, greater risks, and higher costs — will likely diminish during the delay. Genomic science and medicine will still be moving ahead. Thus a ban might at first delay but then bring a swift adoption of the technology, raising the risk that large numbers of people would suffer unforeseen, long-term problems. Paradoxically, the net effect of a policy so cautious that it won't allow any volunteers among us to bear early risks would be to shift greater risks onto the shoulders of our children and grandchildren.

Many critics have claimed that germline manipulation would be foolhardy because complications might show up in some distant generation. Clearly, this is moot with the nonheritable artificial chromosomes I've described, but the argument is weak anyway. Because technology does not stand still, the more generations it takes for a problem to manifest itself, the more likely it is that medicine will have the expertise to deal with it. Moreover, the engineered genome of the first cell of an embryo must successfully orchestrate the entire process of development if that embryo is to mature into a healthy adult capable of reproducing. Almost any genetic abnormality would show up in the first generation, and any serious defect would never reach later generations because it would be lethal or prevent reproduction. A germline intervention that parents judge to be safe enough to use on their child-to-be will be even safer if it can reach their grandchildren.

A different kind of health concern relates to humans as a species. Reshaping our biology might work out well for us individually but bring collective biological problems, diminishing our diversity or robbing us of genetic variants important to our survival.

When people suggest that we should not change our children's genomes because, say, the genetic mutation that brings cystic fibrosis increases resistance to typhoid fever, or the mutation that causes sickle cell anemia raises resistance to malaria, they are asserting that current gene frequencies in the human species are somehow sacred or optimal, which is untrue. This concern is another expression of the admonition that we should not be playing God. Moreover, in the immediate future, the germinal choices of individual parents will have no meaningful effect on the gene pool as a whole, because our immense population — some 6 billion humans — will act as a buffer against such changes.

Cleansing the human genome of particular alleles is not only beyond our ken; once embryos can be screened for disease, no motivation will exist for such a project. But even if we were to drastically reduce the incidence of the gene variant that causes sickle cell disease, no malaria problem would attend this. Germinal choice can take place only in a high-technology world, and long before the sickle cell allele had disappeared through genetic selection or engineering, we would almost certainly have buttressed our medical and public health defenses against malaria. Indeed, in the era of germinal choice, if medical science concluded that having a single sickle cell gene was the best protective against malaria, parents would more likely add than delete the gene.

Political and social disruptions that might accompany germinal choice technologies are a troubling possibility to some. The growing divide between haves and have-nots, for example, is at the center of a snarl of concerns about human germline manipulation. Many people worry that it might bring social inequity, class division, and genetic discrimination that would fragment society. Erik Parens, a bioethicist at the Hastings Center in New York, writes that "the [genetic] lottery could be 'rigged.' The ability to buy not only tools and opportunities to cultivate one's native capacities, but also to buy new or enhanced capacities themselves, would make some individuals doubly-strong competitors for many of life's goods . . . Germline enhancement might widen the already obscene gap between those who have and those who don't."

Parens is right when he says that access to genetic enhancement — or to a sound genetic constitution, for that matter — would be a great advantage. Indeed, precisely this notion — giving a child the best chance in life — is what would inspire parents to use the procedures. Deepened divisions between rich and poor would be only a small part of the resulting story, however, because if germline technology can bring meaningful enhancement, the greatest divisions will not be between rich and poor but between generations.

For all his billions, Bill Gates could not have purchased a single genetic enhancement for his son, Rory. The technology did not exist when he was conceived at the end of 1998. A generation from now, the story will have changed, even if only sophisticated embryo screening is available, but you can bet that any enhancements a billion dollars can buy Rory's child in 2030 will seem crude alongside those available for modest sums in 2060. How society will deal with disparities between rich and poor, much less intergenerational disparities, is unclear, but legal barriers to genetic enhancement will only exacerbate the problem. The rich are always better able to circumvent the law. They, not the poor, can travel to wherever enhancement is legal, buy the desired genetic procedure, and pay to evade prying tests back home.

A similar situation existed prior to the Supreme Court's 1973 *Roe v. Wade* decision, which granted women the right to a first-trimester abortion. Affluent Americans had relatively easy access to safe abortions; the poor did not. More recently, in 2002 wealthy Germans could obtain preimplantation genetic diagnosis in Brussels or London, despite its illegality at home. The challenge facing those who, like Parens, worry about unequal access to these technologies isn't how to ban them, but how to ensure that they become cheap enough for everyone to afford.

An important undercurrent in discussions about the selection or manipulation of embryos is fear of genetic discrimination. People worry that they could lose their jobs or health insurance if some genetic vulnerability of theirs is identified. Similar concerns apply to our children. The question is how society will treat those with genetic imperfections once parents have the power to prevent them.

Look at the present possibilities of genetic discrimination, dangers that have nothing to do with genetic manipulation and everything to do with genetic testing. Even today, DNA chips can screen for tens of thousands of genetic variants at a time, so routine screening will soon be a reality. Will this information be kept private? Government regulations now exist to oversee the institutional use of our medical and genetic records, but that doesn't mean that we will be able to keep our genetic profiles out of the hands of someone who really wants to obtain them. We all leave a trail of biological debris behind — dandruff, saliva on toothpicks and napkins, stray cells on Q-Tips, fingerprints. Evolution has had no reason to guard our DNA from prying eyes, and it doesn't.

I suspect, though, that genetic discrimination based on DNA chip profiles will not be nearly the problem imagined. The main reason is simple. The things we most want to know to guide us in our dealings with people are the traits that we see or can readily probe: temperament, ability, appearance, intelligence, age, marital status, philosophy, gender, reliability, friendliness. We don't need to look at genes to evaluate people or discriminate against them. And as for outright racism, some might fear that genetic testing will be used as a way to identify individuals who do not have the stereotypical appearance of a particular ethnic or racial group but are connected to it by ancestry. In some places, the use of genetic testing might lead to more pernicious forms of racism, but I suspect that more often our increased genetic knowledge will break down racism, by showing people that their family lineages are much more mixed than they might have thought.

Dateline, 2015: A job interviewer sits at her desk quietly scrutinizing the graphs and bar charts on the genetic profile of the applicant seated across from her. A frown flickers over her face as she realizes that the applicant just doesn't have the right genes for the position. The profile clearly shows that the person is prone to obesity and baldness, is probably unattractive, and has a chromosomal structure indicating the wrong gender. This is a man! Genetic tests do not lie; the applicant is not the topless waitress she's looking for. After a few comments, she shows the paunchy, balding man the door. As he enters the elevator, he dejectedly mutters that genetic discrimination has ruined his chances again.

Absurd? Of course. But some of our fears of a future *Gattaca*-like era of genetic discrimination aren't much more realistic. Most of the common reasons for discrimination — age, gender, race, ethnicity, weight — are obvious at a glance. Bad hygiene or obnoxiousness will turn someone off far more than a genetic printout ever will. Evolution has primed us to respond emotionally to phenotype, not genotype. In most situations, genetic predispositions will never be as useful as talking to people, looking at their past performance, testing their abilities, and phoning their references.

Disease susceptibilities are another matter. Some argue that companies would be unwilling to hire a person with susceptibility genes that raise group insurance rates. This is no doubt true, and in the United States such problems have already arisen. The fundamental challenges that genetic testing will present to the current structure and goals of health insurance, however, are better handled by regulating insurers and employers rather than biotechnology.

Insurance is a mechanism for people to share unknown risks, so as we each identify ever more of our risks, insurance will become ever less useful to us. An obvious way to keep insurers from penalizing people at elevated risk is to provide a guarantee of genetic privacy, and efforts are under way to accomplish this. But such privacy creates an information imbalance with its own problems. If people can test themselves to find out if they are at high risk, and load up on low-cost life insurance or top-of-the-line health insurance, they will. Notwithstanding the personal tragedies embodied in these situations, better ways than institutionalized insurance deception must exist for caring for such people. Moreover, as testing technology continues to improve, such an arrangement breaks down. Claims mount, rates climb for healthy individuals, and those who can show they're at low risk will obtain special deals. Eventually, insurance for people at high risk becomes as expensive as it would have been without genetic privacy.

The challenges accompanying genetic testing will spill over into the reproductive arena as PGD evolves and the birth of a child with serious genetic infirmities or vulnerabilities becomes avoidable. One of the more troubling arguments against letting parents choose the genes of their future children is that such control would so diminish the numbers of children with genetic diseases that society would begin to ignore those still afflicted. This concern is easy to understand, but by its logic, we should stop medical treatment of ill adults as well. After all, if cured, they might be less committed advocates for others with their condition. But I suspect that this fear is exaggerated. We have limited resources, so reducing the number of people with diseases and disabilities might bring those remaining more, not less, help. Rarer diseases do tend to receive less research funding, but overall, this makes good sense. Progress on common diseases helps more people and frequently speeds the development of treatments for less prevalent disorders. If we first tackle common, tractable illnesses, we can later apply what we learn to rare and resistant ones.

Some bioethicists assert that germinal choice will corrupt the attitudes of parents toward their children and affect children's self-images. But these are merely conjectures. Some parents will obviously be disappointed if their "designer" kids don't turn out as hoped, but no one can say how common such disappointment would be, whether it would persist, or if it would be as prevalent as that which already occurs today. People have children for many reasons and start out with a variety of expectations. Some fathers and mothers are happy no matter what their kids do; others doggedly push them toward predetermined goals. Mate selection is a potent force in shaping our offspring, and we devote much attention to this choice. When we can do even more to influence our children's genetics, we will struggle with these decisions too. Future sources of parental dissatisfaction are easy to predict. Some parents will forgo germinal choice technology and end up wishing they had used it. Others will use it and be disappointed in the results.

The same uncertainties will apply to a child's image of his or her own destiny. Offspring do not arrive with a clean slate; they carry the hopes and fears of their parents. Children's temperaments, circumstances, and ongoing experiences circumscribe their choices. In the future, genetic tests will tell kids about their genetic vulnerabilities and predispositions at an early age, and this information is bound to influence them whether or not their parents chose their genes.

At this point, no one knows whether this knowledge will help guide young people during the often painful period of self-discovery and exploration we all pass through or will squelch their early hopes and dreams, channeling them in unfortunate ways. Nor can we know what they will feel when they learn that parental choices shaped them, not chance or the "hand of God."

A friend of mine who was adopted at an early age said she was happy when her adoptive parents told her as a little girl that they had specially chosen her from all the babies they looked at. To learn that her parents had instead flipped a coin would have been no gift. Perhaps future "designer children" will feel like winners from birth, because they will not have that suspicion that might lurk at the back of our minds when we think of germinal choice. Each of us knows that had embryo screening been in use at the time of our conception, our parents might have chosen another, "better" embryo in our place.

Like issues of religion and morality, no consensus will emerge about how genetically selecting and manipulating children will affect families. People's attitudes — as is the case with ideas on the best way to raise a child — will grow out of the differing relationships and experiences they have with their own parents and children. Some resolution of the question may come with the arrival of real children in real families, but I suspect that even this won't tell us much. In any study, controlling for inherent biases will be too difficult.

Parents who choose these technologies, particularly those who are the first to adopt them, will have temperaments and beliefs that differ from more cautious parents who don't. Children conceived using GCT will, on average, differ from those who weren't conceived by parental selection. And early children of GCT may be treated differently by their peers and siblings. Even if obvious differences in family dynamics appear, teasing apart the various influences will be hard. And if we see no differences, we still won't be able to say that psychological problems won't surface if we wait longer.

In short, while agreement about the effects of germinal choice on family dynamics and children's psyches may not be as difficult to achieve as consensus on religious issues, it will come very slowly. Moreover, there is a logical flaw in the assertion that germinal choice technology will damage children. Because such children would not exist without the technology, those who are alive by virtue of it cannot be its victims unless their lives are so piteous that somehow they would be better off not to have lived at all.

We also have to wonder about the spiritual consequences if we modify our biology. Perhaps an extended human lifespan would change the rhythms of our lives or shift our values in undesirable ways. Genetic selection might leave us purposeless and adrift or entice us to seek what we perceive as human "perfection." Maybe the hubris behind these interventions would somehow corrupt our spirit. Such vexing notions, however, could apply equally well to artificial intelligence, nanotechnology, and other developments now surging forward. Whether or not we alter our biology, our world will change dramatically over the next century. These changes will deeply challenge our values and beliefs. They will bring us losses as well as gains.

The possibility of spiritual decay has been raised many times in many eras. The industrial revolution, global warfare, labor unions, income taxes, universal education, contraception, television, the computer, women's rights, all have brought dramatic shifts in the way we see ourselves. All threatened the status quo. Our world is far from perfect, but if we each could somehow choose to leave the present behind and return to a past era stripped of the romanticism time brings, few of us would go, and I suspect that most who did go would come to regret their choice.

Humans adapt. Even if the future turns out to be one that we ourselves would find as foreign and troubling as our great-great-grandparents might find our day, those who come of age in the future will likely find it appealing. A few romantics may one day look back on our era as a golden one, but far more future humans will see today as a primitive, difficult time, far inferior to the world they know.

Nor will new technology erase our flaws or take away the foibles and imperfections that add meaning to people's lives. For every difficulty these technologies solve, they will bring a new problem to occupy us or will raise our expectations, so that what seemed insignificant begins to disturb us. By virtually every material measure, we in the developed world have never had it so good, yet many feel that our problems have merely multiplied. If we want to gauge how far we have come in the century just past, we need to look at how much people's expectations have risen. Today we expect not just to survive but to be fulfilled, to live out long lives, and to grow old in good health.

Nothing is wrong with arguing that the advent of human biological manipulation will cause us to drift from our spiritual moorings, but to use this argument as a foundation for public policy would be a grave mistake. Like the other issues of morality and religion mentioned earlier, no amount of discussion will resolve questions of spiritual malaise.

* * *

"Slippery slope" is an umbrella term for all our general concerns about coming reproductive technologies. Some bioethicists have argued that once we start down the path of human biological manipulation and germinal choice, we will be unable to turn back. The first steps may seem reasonable, even beneficial, they say, albeit often rhetorically, but one thing will lead to another, and soon we will be changing our genetics in ways we would never have dreamed of. The only protection against the most egregious of imagined possibilities is to draw a line right at the outset and not cross it. George Will asserted just this in an opinion piece: "Positive eugenics, any tailoring of an individual's genetic endowment . . . ," he wrote, "will put us on a slippery slope to the abolition of man."

The challenge in refuting this argument is that it is so, well, slippery. It requires no evidence of immediate danger and is not weakened by refutations of any specific hypothetical threat. Conjuring up grim futures is easy, and the metaphor of the slippery slope has been used time and again to oppose all kinds of innovations. But if biological manipulation is indeed a slippery slope, then we are already sliding down that slope now and may as well enjoy the ride. After all, we already use birth control, *in vitro* fertilization, and preimplantation genetic diagnosis. We already clone sheep, manipulate mouse genetics, and alter human genes to fight disease.

If we can make choices about technology today, and I believe we can, we will be able to do so in the future. Technology doesn't emerge magically; it depends on the existence of large numbers of people who want it. Today we are actively choosing the technologies that serve us, and if future generations do the same, people's biggest fears will not come to pass.

To continue the metaphor, advanced reproductive technologies are more like a slippery sidewalk than a slippery slope. Rather than sliding uncontrollably into some deep abyss, we more likely will take a spill or two, get up, brush ourselves off, and continue cautiously on our way. Given the difficulties of broadly implementing these technologies, and given our disinclination to injure our children, it is hard to come up with believable scenarios in which foolhardy use of these technologies would persist long enough to be as big a health hazard as alcohol, cigarettes, automobiles, lack of exercise, or poor diet.

Since germinal choice technology has already arrived, the question is how we will deal with its growing power. A promising approach would be to concentrate on the specific choices parents can make about their children's genes and try to discourage those practices that seem damaging. Our current approach, however, is to worry about the technicalities of the procedure itself: which tissue a cell comes from, when and how it is collected, what techniques are used to manipulate it. Paradoxes and contradictions will forever frustrate this method, because it hinges on philosophical abstractions far removed from dangers to present or future humans.

But as GCT advances, it will become more real, because it will hit closer to home. Watching some fellow on television talk about human cloning or an embryo test for a rare disease you've never heard of is not the same as learning that your daughter is thinking about choosing the genetics of your grandchild. And endless debate over whether we should be playing God or whether advanced GCT will corrupt us spiritually will not get us far. Nor will arguing about the unknowable future consequences for children or parents or society.

The areas that would most benefit from discussion right now are concrete matters: how to test these technologies, appraise their risks, monitor research, and minimize clinical abuse. We also would do well to consider how various policies will affect patterns of access and use, because the real question before us is not whether these technologies will appear, but when they will, who will have access to them, and how we will use them.

8

The Battle for the Future

How good bad music and bad reasons sound when we are marching into battle against an enemy.

— Friedrich Nietzsche, 1881

AS ADVANCES in genomics and *in vitro* fertilization unite to bring us such technologies as germline manipulation and in-depth embryo diagnosis, must there be a battle over their use? Policy-makers might, after all, acknowledge the arrival of these technologies, accept that people differ in their attitudes toward them, realize that society will adjust as it has to past advances such as the birth control pill, and support efforts to minimize risks and maximize benefits.

Unfortunately, that scenario is unlikely. So symbolic and evocative of people's fears is the manipulation of human embryos that many countries have already banned the procedures. In Germany, either germline manipulations or preimplantation genetic diagnosis can bring a five-year prison sentence, according to the 1991 Embryo Protection Law (now being reconsidered), and the minister of justice stated that the purpose of the law was to "exclude even the slightest chance for programs aimed at so-called improvement of humans."

Even in the relatively permissive United States, moves are under way to hold such research apart and subject it to special scrutiny. In

September 2000, a committee under the auspices of the American Association for the Advancement of Science (AAAS) urged an immediate block on a wide range of clinical procedures that the group labeled "inheritable genetic modifications" (IGM). These included reproductive cloning, germline procedures, and fetal therapies that might alter eggs or sperm. Because cellular fluid, called cytoplasm, contains about a dozen special genes that are inherited, the committee even recommended banning cytoplasmic transfers into the eggs of women suffering from a cellular disorder that would otherwise keep them from having children.

> Inheritable genetic modifications cannot be carried out safely and responsibly on humans utilizing current methods . . . Even if we have the technical ability to proceed, however, we would need to determine whether IGM would offer a socially, ethically, and theologically acceptable alternative to other technologies . . . Until then, no research or applications that could cause inheritable modifications in humans should go forward.

No one would dispute the observation that direct germline modification is not safe today. But when committees that include scientists suggest that such procedures need to be "theologically acceptable" before their use, something unusual is going on. Fetal gene therapy aligns well with other gene-therapy work. Cytoplasmic transfer seems safe and is an obvious extension of other infertility research. No one is presently planning germline interventions of this sort, and in any event, no more than an occasional ill-conceived procedure — probably no worse than countless others with less provocative intentions — could slip by the review boards that oversee medical research and practice.

With mechanisms already in place to discourage doctors from trying risky, speculative, and unwarranted procedures on people, the worry about germline modification appears unnecessary. Once the procedures become more feasible, direct oversight is bound to follow, if only because without it, few couples would have the nerve to use them and few providers would be willing to shoulder the liability risks. In short, the inherited genetic modifications that the AAAS committee refer to are not about to explode into clinical use.

No useful interventions exist yet. No gene transfer procedures are available. And when these interventions and procedures finally do exist, most parents will want evidence that they are reliable – validations that would take many years. Direct genetic modifications are so unlikely to touch large numbers of pregnancies in the next decade that anyone wishing to protect babies would do better to focus on poor nutrition, alcoholism, and unsafe drinking water.

The major threats from coming advances in reproductive technology are not medical but political, social, and philosophical. Moreover, embryo screening, which is safe, legal, and already in use, will soon bring up the same troubling dilemmas of germinal choice that the AAAS committee hoped a ban on inherited genetic modifications might delay. Emotions surrounding human genetic selection and manipulation are still relatively subdued compared with the passions that abortion now evokes, but to most people, "designer children" seem distant.

The current discussion about human enhancement is not what it seems, however. It is not about medical safety, the well-being of children, or protecting the human gene pool. At a fundamental level, it is about philosophy and religion. It is about what it means to be human, about our vision of the human future.

Leon Kass expressed his revulsion with cloning, but he could as easily have been referring to any advanced germinal choice technology (or perhaps to *in vitro* fertilization, which he assailed twenty years ago). "We are repelled . . . because we intuit and we feel, immediately and without argument, the violation of things that we hold rightfully dear," Kass wrote. "We sense that cloning represents a profound defilement of our given nature as procreative beings, and of the social relations built on this natural ground . . . Repugnance may be the only voice left that speaks up to defend the central core of our humanity. Shallow are the souls that have forgotten how to shudder."

Who Will the Early Enhancers Be?

The specter of forced sterilization, concentration camps, and other horrors of the twentieth century perpetrated under cover of eugenic rhetoric ensures that today virtually no one speaks about the possibilities of human genetic manipulation without caveats and disclaimers. Many would go beyond such cautions, however, and reject this realm entirely. This opposition ranges from religious conservatives who see embryo research as tantamount to human experimentation, to bioethicists attuned to the potential problems of new technologies, to scientists troubled by the power that genomic sciences will wield, to environmentalists and neo-Luddites worried about so deep an infusion of technology into our lives.

But many others say they would use the reproductive technologies if they existed. In the international opinion poll mentioned earlier, from 22 to 83 percent of those surveyed in each of eight countries said they would use safe genetic interventions to enhance the mental or physical attributes of their children. Between 62 and 91 percent said they would use gene therapy to keep a child of theirs from inheriting a disease like diabetes.

Almost no one, however, is pushing for these technologies; they are too threatening. I once spoke about these issues with a group of libertarians, who rail about taxes and wholeheartedly embrace global free markets. Even they voiced concerns about unregulated manipulation of human embryos. Clearly, human biological enhancement puts philosophies of individual autonomy and laissez-faire ideology to the test.

The zealous few who most support these technologies are those who want to use them personally. Cloning is an instructive example of what will occur on a much broader front as other germinal choice technologies evolve and mature. Following the birth announcement of Dolly, the famous cloned sheep, the Raelians, a New Age religious group that believes that visitors from space spawned humanity, were among the first to push aggressively for human cloning. In the fall of 2000, they announced that a wealthy American had donated $500,000 to fund a cloning attempt, and that they would move ahead. When I spoke with the church's founder, Raël, a Frenchman previously known as Claude Vorilhon, he told me that aliens had visited him in 1973 and shown him machines that could clone humans and sprout them into full-grown adults. Impressed, he decided to become their messenger, and raise $20 million to build an embassy to house the extraterrestrial visitors who would one day bring our salvation.

The Raelian cloning project was sufficiently quirky to command instant media attention. And Raël not only had money, he had some fifty volunteers willing to serve as surrogate mothers. With these resources, the group may succeed once mainstream researchers surmount the technical problems of nuclear transfer in primates and overcome the tendency of cloned embryos to have the unpredictable gene expression that can disrupt placenta formation and lead to fetal abnormalities.

My point here is that as reproductive technology progresses, it is bound to drift out of the hands of traditional medical researchers and clinicians. But cloning efforts in more mainstream quarters are also moving forward. When Panos Zavos, a reproductive physiologist at the University of Kentucky, announced (shortly after the Raelians did) that he and an Italian fertility doctor, Severino Antinori, would clone a human, the U.S. Congress decided to hold hearings. The testimony about the medical dangers of proceeding with human cloning persuaded neither the white-robed Raël nor Zavos in his suit and tie to shelve their projects. Instead, Zavos asserted at the hearings that he would not operate on American soil and would be unaffected by any U.S. ban, and the Raelians said that they too would simply take their project elsewhere.

Whether either effort will get off the ground, much less succeed, is uncertain, but the ultimate outcome is not. Once science progresses further and animal cloning procedures become safe and reliable, there will be little delay in demonstrating them in human volunteers. The global regulatory environment will likely determine where such attempts take place and whether they are public, but not whether human cloning will occur. The same will also be the case with other advanced germinal choice technologies.

The Raelian episode illustrates that when people start to use these technologies, odd combinations of high technology, religion, and other beliefs may appear. I had heard several anecdotes, for example, about women using *in vitro* fertilization merely to time their pregnancies, so in the fall of 2000, when I spoke in San Diego before the American Society of Reproductive Medicine, I asked the large audience if anyone had encountered such behavior. A Brazilian IVF specialist volunteered that one of his patients had been so involved with numerology and astrology that she had decided to use IVF to fix the time of her child's conception. Here was a case of using reproductive technology to achieve an occult rather than a genetic enhancement.

Those most committed to the goal of human enhancement are probably the so-called transhumanists, a hodgepodge of individuals and organizations loosely united by a desire to transcend human limitations. They welcome the development of anti-aging medicines, smart drugs, and genetic modification. The Extropians — a group studded with bright, iconoclastic figures and a board of directors that includes Marvin Minsky, Ray Kurzweil, and Roy Walford — are active in this arena. In 1999, their annual meeting in Berkeley, California, included discussions focused on the challenges of extending human lifespan and genetic engineering. Walford, Judy Campisi, Calvin Harley, and Cynthia Kenyon, well-known figures in the field of aging, spoke at the meeting. At the conclusion, Max More, who founded the Extropians (the name is meant to signify the reverse of entropy, the universal trend toward disorder) in 1992, read a "letter to Mother Nature," which captures their attitude:

> Mother Nature, truly we are grateful for what you have made us. No doubt you did the best you could. However, with all due respect, we must say that you have in many ways done a poor job with the human constitution. You have made us vulnerable to disease and damage. You compel us to age and die — just as we're beginning to attain wisdom. And, you forgot to give us the operating manual for ourselves! . . . What you have made is glorious, yet deeply flawed . . . We have decided that it is time to amend the human constitution . . . We do not do this lightly, carelessly, or disrespectfully, but cautiously, intelligently, and in pursuit of excellence . . . Over the coming decades we will pursue a series of changes to our own constitution . . . We will no longer tolerate the tyranny of aging and death . . . We will expand our

> perceptual range . . . improve on our neural organization and capacity . . . reshape our motivational patterns and emotional responses . . . take charge over our genetic programming and achieve mastery over our biological and neurological processes.

This image of the human journey toward a superior "posthuman" may be difficult for many to take seriously, but the determination to use whatever new technologies emerge from today's explorations of human biology aligns well with prevailing attitudes. The number of people who undergo cosmetic surgery or pop expensive vitamin boosters testifies to this. Whatever people's philosophies of human enhancement, their decisions about using specific procedures often hinge on cost, safety, and efficacy rather than political or social consequences.

Understanding the forces that will affect the adoption of today's budding germinal choice technologies requires a global perspective. After all, outside the West, some scientists will push ahead regardless of what happens in North America and Europe. Societies in the Middle East and elsewhere that are driven by religious ideology will hardly embrace these innovations, but they will not figure heavily in the equation of global change. The fleetest, not the most cautious, will set the pace.

China stands out among today's emerging nations because it seems to have the resources, the predisposition, and the self-reliance to independently pursue the new technologies. Whether such a project is on China's agenda is unclear, but Beijing has been aggressive in managing reproduction, and the social imperatives that grip the country seem to be pushing it in this direction. The one-child-per-family policies that cut its population growth from 2.9 percent in 1970 to 1.4 percent in 1990 were an extraordinary accomplishment. China's 1995 Maternal and Infant Health Care Law, which calls for compulsory premarital checkups as well as sterilization for "genetic diseases of a serious nature" and "mental diseases," is not now enforced, but its passage in the face of a loud international outcry is evidence of the country's willingness to follow an independent course. And in its rapid embrace of genetically modified crops, China has shrugged off European fears and released over one hundred varieties, more than any other country, an indication of a growing enthusiasm for genomic technologies. Its total production of genetically modified crops is still small compared with that of the United States, Canada, and Argentina, but it is racing to catch up.

The same also seems to be happening with reproductive technologies. The IVF clinic in the city of Xi'an is not at all in accord with Western stereotypes about China. With its double containment entrance, gleaming equipment, and scurrying masked nurses, the laboratory looks more like a set from *The Andromeda Strain* than an IVF clinic. The evocative air-lock entry is merely part of a filtration system needed to cleanse the polluted air hanging over the city. But this initial impression brings up the obvious question: why is China, with its teeming population, its aggressive control of family size, and its push for rapid economic growth, so concerned about infertility?

The army of terracotta soldiers that draws so many tourists to Xi'an interests me less than the giant billboards in Beijing's Tiananmen Square advertising infant formula. Overlaying huge images of Einstein and Picasso are slogans suggesting that milk may make your child more like these geniuses. Perhaps similar ads one day will tout the benefits of genetic enhancements. Some forty IVF clinics already exist in China, often built with military assistance, and a commitment to such technology seems to exist in influential segments of the government and military.

Many in the West shake their heads over Chinese eugenics and sterilization laws. But where couples can legally have only one child and must look to that child as their primary provider in old age, they are bound to feel strongly about having the "best" child they can. If this means having a boy — because a girl marries and joins her husband's family — they will try to have a boy. If it means screening for genetic vulnerabilities and diseases, they will do that. And if they could make their children smarter or stronger through genetic testing and manipulation, they'd probably do that too. Although the Chinese government may become involved in orchestrating parental "choices" of germinal enhancement, no directives would be needed for the technology to thrive in China, because the coming possibilities accord so well with the perceived self-interest of the populace.

Wouldn't parents in the West, given similar social constraints, feel the same way? After all, the option of caring for a seriously disabled child or passing that responsibility on to the state is a modern luxury that few possess. Polls suggest that even with this option, 80 percent of Americans would use genetic interventions to prevent a child from inheriting a fatal disease. I suspect that couples with small families will be particularly inclined to screen the genetics of their future children carefully.

The counseling that medical geneticists give to prospective parents offers another window into the differing attitudes found in different cultures. If a prenatal test indicates the presence of Klinefelter syndrome, the most common form of dwarfism, 92 percent of counselors in China would push the parents to terminate the pregnancy, whereas in the United States, Australia, and most of western Europe, fewer than 10 percent would. The figures are similar for some two dozen diagnosable conditions ranging from Down syndrome to cleft palate; nearly 80 percent of Chinese counselors would urge abortion.

In more than half the prenatal tests considered, Chinese geneticists counsel abortion more often than their counterparts anywhere else, though in India, Russia, and a number of smaller countries, including Greece, Cuba, Turkey, and Hungary, counselors also routinely give that advice. More than 90 percent of these counselors — compared with fewer than 20 percent in northern Europe and the United States — feel that to knowingly bring into the world an infant with a serious genetic disorder is socially irresponsible. In countries with a combined population of more than 3 billion — that is, a majority of all humans — most geneticists say the eugenic goal of reducing the number of deleterious genes in the population is an important one.

The Threat of Human Enhancement

A future in which parents select meaningful aspects of their children's genetics engenders anxiety in many people, despite the fact that our selection of a mate has always included such reckonings. Perhaps the use of technology for this purpose seems too controlled and calculated. To see why even people with no strong religious objections to advanced germinal choice technology might want to prevent it, imagine the following best-case scenario: Medical geneticists learn how to substantially reduce our vulnerabilities, improve our health, extend our vitality and lifespan, and enhance various other human attributes. Complications turn out to be modest and manageable. Public opposition, strong at first, soon withers into insignificance. Parents are thoughtful about the choices they make for their future children. The technology is sufficiently simple and inexpensive to be widely available. Totalitarian regimes don't try to impose it on their populations.

Although this scenario brims with utopian benefits and embodies a dreamy perfection that ignores the inherent messiness of any development so profound, it would leave extraordinary loss in its wake. Such a course of events would not only create a gulf between generations, it might divide us more deeply by encouraging us to judge explicitly the value of various human attributes. GCT interventions would so change the trajectory of human life and the essence of the human condition as to render our past irrelevant to us in many ways. In making ourselves anew, we would have to figure out all over again what was important to us, and how and where we would find meaning in our lives. These are not easy tasks today, but at least the terrain is familiar and we can draw on history, religion, and literature for guidance.

As we head down the path of biological modification, we will gradually cease to be who we have always been. If we live longer, healthier, vastly different lives, we may end up estranged from the world we now inhabit. We may cease to feel connected to humanity as a whole. Such possibilities are why some people so vehemently oppose the new reproductive technologies. To them, germinal choice will bring the invasion of the inhuman, the displacement of the born by the made, and the twilight of humanity.

Dr. Nigel M. de S. Cameron, professor of theology and culture at Trinity International University in Illinois, suggested as much when he argued, in May 2001, for an outright ban on the cloning of human embryos, even if they would be used for medical rather than reproductive purposes: "Cloning a human baby isn't just bad, or unfortunate, but something which would be profoundly evil because it would constitute a new human being in a radically defiled and deformed moral fashion. That is the view that many, many people take. It seems to me it is not a religious view. It's a view coming intuitively out of our vision for human dignity."

Such a reaction could not be further from those of the Extropians and others who eagerly welcome the emerging possibilities and see them as the flowering of humanity, the realization of what until now has been an unreachable human dream: transcending our biological limits. The chasm separating these two perspectives is at the heart of the coming struggle over our journey into the human future.

Fear about the huge changes that may attend our manipulation of human biology and reproduction is understandable, but blocking advanced GCT will not protect us. Change is arriving too quickly across too broad a front. Our accelerating disengagement from the lessons and truths of the past will not be easy. We face it already in our shifting cultural environment: the weakening of traditional institutions and values, our multiplying connections with distant others, the increasing seductions of electronic worlds, the growing responsiveness of the digital devices we are infusing with intelligence and language. Germinal choice technologies will be but one ingredient, albeit a critical one, in this tumultuous transformation of human life.

There is little doubt that germinal choice will eventually reach us, but the extent of the divisiveness it brings is still uncertain. The harshest conflicts will probably arise not between societies but within them. International disagreements about manipulating human embryos may generate impassioned rhetoric, but once nations that oppose such technology enact bans within their own borders, they will be able to do little more. Nations have never been very successful in persuading other nations to modify their views on religious and social matters, and GCT will be no different. Even an international ban on these technologies would have no lasting impact since broad public interest in them would preclude effective enforcement. The more widespread and aggressive the opposition to their use, the greater the financial incentives will be for individuals and nations to defect and cater to pent-up demand.

Let's return to China, which was the preeminent world power in 1500 and might imagine itself achieving that position again in a few centuries. If the manipulation of human genetics seems a necessary step along that path, Western sensitivities and policies are unlikely to stand in the way. And once a single major nation embraces so foundational a development as this, others would soon have to follow, however reluctantly, to avoid being left behind.

In societies where culture and religion are relatively homogeneous, internal conflicts may be moderate, because opinions about these matters will be less diverse, but divisive policy battles are a virtual certainty in multicultural cauldrons like the United States. Some twenty-five years after *Roe v. Wade* the United States still experiences occasional bombings of abortion clinics.

Human enhancement procedures will bring an even more traumatic struggle. The debate, though polite today, will become much harsher as sophisticated embryo selection and germline technology draw near. Public reaction to the first human cloning may provide a foretaste of the passions that will be unleashed, but given the developmental abnormalities that can arise from gene expression errors caused by these procedures, human cloning may not happen nearly as soon as enthusiasts have predicted. In any event, those who imagine that public opposition will melt away with the arrival of a cute, healthy baby are underestimating the strength of people's feelings about cloning and GCT in general.

Of course, prior developments might moderate people's attitudes. If somatic gene therapy and the genomic sciences start to deliver solid medical breakthroughs, fears may diminish. Our pets may play a role in shaping public opinion as well. Genetic Savings & Clone and other companies have been working to clone dogs and cats as a commercial service. If this turns out to be relatively safe and reliable, and the wealthy begin to clone much-loved older pets to make a twin puppy or kitten, we may grow comfortable with these reproductive technologies sooner than we think. But I doubt that the coming general transformation of human reproduction will be so easy. The Catholic Church, after all, still opposes birth control.

The irony about germline and other germinal choice technologies is that as challenging as they will be at a symbolic level, the procedures are fundamentally life-giving. This could pose a dilemma for those who might be tempted to smash the laboratories where such work takes place. Today anti-abortion extremists see themselves as locked in a battle to save innocent lives from the knives of murderers. They assault abortion doctors and display gruesome photos of aborted fetuses in a crusade they see as "pro-life."

Advanced reproductive technologies will present more complex choices. Genetic alteration of human embryos may be deeply offensive to these crusaders, but they would see an incubator filled with embryos as a vessel holding human life. If life begins at conception and is to be honored as such, to destroy that incubator would be tantamount to mass murder. Indeed, some who vehemently oppose abortion may find themselves drawn to the idea of practicing medicine on embryos, simply because it seems to treat these cells as patients.

Protecting the Human Race

Those who categorically oppose future reproductive technologies such as cloning, advanced preimplantation genetic diagnosis, and germline enhancement portray themselves as voices of reason and caution, trying to protect humanity from these divisive and dangerous spinoffs of medical science. Whether they depict the threat as medical, social, environmental, or spiritual, their implicit assumption is that if we cannot stop these technologies, we should at least delay them as long as possible, because of the significant risks they pose and the dubious benefits they offer. Given that humanity has not needed such novel reproductive procedures to reach the present and has already suffered lamentable forays into eugenics, they see no good argument for pushing ahead.

There was a furor when it was revealed in December 1999 that the European Patent Office had inadvertently awarded a patent dealing in part with human germline manipulation. The European Parliament quickly called for the patent's revocation and reiterated its opposition to applying to human beings biotechnology that involves "interventions in the human germline," "cloning of the human being in all phases of its development," or "research on human embryos, which destroys the embryo."

Serious enforcement of these categorical prohibitions would involve two costs: the beneficial enhancements that are lost and the damage the ban itself inflicts. The first loss might be relatively minor, because research can usually migrate elsewhere. But the extreme measures needed to effect such a ban might be quite injurious. What's more, the larger goals of such actions are murky at best. If these bans are a call for "natural" reproduction, unpolluted by technology, then why not also repudiate IVF and birth control? If they are a declaration against eugenic selection, then they leave gaping loopholes such as amniocentesis. If they are meant to protect distant, yet-to-be-conceived generations, then they are so blunt and premature that they could do more harm than good.

Advanced germinal choice technology does not yet exist, so enforcement is not an issue, and prohibitions are easy political gestures. But once GCT arrives, enforcement will be nearly impossible. Testing every baby at birth, for example, could reveal enhancements involving artificial chromosomes but not single-gene replacements. Even running genetic profiles on parents would not show that a child was from a genetically screened embryo or was the clone of a deceased sibling, since the child's genome would not be altered.

Many people would be more disturbed by mandatory genetic tests of babies than by the chance that some couple somewhere might enhance a child. The International Olympic Committee had to discontinue testing the gender of female athletes after the practice was assailed as an invasion of privacy, and in the United States, the privacy provisions of the Health Insurance Portability and Accountability Act make the sharing of any medical information without a patient's permission, including genetic information, a crime.

Punishment would present problems too. A baby born by germinal choice technology would be guiltless, penalizing the parents harshly enough to deter them from a private act intended to help their child might seem abusive, and reproductive specialists might be beyond legal reach. Not only does the diversity of attitudes throughout the world suggest that any serious ban could only be regional, but in the United States, much GCT may eventually be interpreted as procreative freedom and treated as a fundamental right protected against unwarranted government intervention. The ultimate effect of regional bans will be to move these technologies to permissive locales and cede their development to others.

If we ban advanced GCT on philosophical or religious grounds and cannot enforce the ban, it will cause harm in other ways. Prohibitions would lead people to judge others not by who they are but by the way their parents conceived them, to label such children as different, and to invite the very class conflicts we wish to avoid. This course would ensure that advanced reproductive technologies besiege rather than serve us. It would enshrine in law the fear that we cannot trust parents to use these tools responsibly. It would push us toward government monitoring of our genetics and reproduction.

Even if we could ban all germinal choice technologies, we still would not avoid grappling with the consequences of unraveling human biology. Our growing ability to read the genetics of potential future children and make choices based on what we learn is but one way that the genomics revolution will challenge our sense of who we are, how we relate to one another, and what is important to us. With the cost of DNA chips falling rapidly, routine analysis of our individual genetic constitutions is not far away. Like it or not, we will have to come to grips with what our genes have to say about who we are. We will have to face how our genetics circumscribes our potentials, our vulnerabilities, and even our personalities. Our genes will not tell us our destiny, but they will speak to us, and we would do well to listen.

We can pretend that every genetic makeup is as good as every other, but putting aside serious genetic diseases and vulnerabilities, we each have our likes and dislikes, our prejudices and preferences.

As individuals, we do not respond alike to all personalities, and when we can ascertain something meaningful about the genetic predispositions of our children-to-be, we will likely be swayed by the information. Some parents will not want to know such things, just as they do not want to know the sex of their child in advance. Others will be curious and want a look. Still others will want to make choices.

We might assert that knowledge is okay but intervention is not, though this would be contrary to current practice and law in the United States. Couples already make some reproductive interventions — such as screenings for disease — that are widely accepted and others — such as sex selection — that are not. As the quality and quantity of information grow, a consensus will emerge about some practices, but for many others no consensus will be possible. If we decide to impose regulations on the more contentious choices, we will be heading down a difficult path. The law is a blunt instrument when it comes to the nuances of individual situations, and it will seem ever more out of touch as our choices become subtler. Imagine having to seek permission for the timing of your child's birth, laying out your reasoning and pleading your case. The most even-handed committee would seem oppressive, because the process would be so intrusive. I suspect we would do better to rely on parental decisions, unless consensus exists that there is a likelihood of serious harm.

This hands-off approach would require great restraint, because many situations will no doubt distress us. Some deaf couples say that in their culture a deaf child fits in more easily and therefore would be better off than a hearing child. If such a couple decided to use embryo selection to guarantee that their child was deaf — and some have expressed that wish — no one could argue that they were injuring a healthy child. They simply would be choosing deafness.

Testing early-stage fetuses and aborting those not deaf may be disturbing, but so strong is the belief in parental autonomy in the United States that more than a third of American obstetricians say they would perform a prenatal test to open up that possibility. Prenatal tests for sex selection are even more accepted. Given that we do not prohibit these controversial procedures, a general ban on preimplantation genetic diagnosis of embryos seems unlikely, and telling women which embryo to implant is implausible. Indeed, once broad embryo selection is available, even a ban on direct germline manipulation is dubious because the technique's intended outcomes would be so obvious an extension of embryo selection.

Sports offers another preview of the challenges of enforcing any ban on genetic modifications. The use of performance-enhancing drugs is illegal in competitive sports, but policing is difficult despite a nearly ideal situation for effective action. The sanctions enjoy near-universal public support. A straightforward penalty — exclusion from the sport — exists. Sporting authorities have large budgets at their disposal. Athletes submit to intrusions on their privacy that the general population would never tolerate. Elite athletes are few in number, easily identifiable, and physically accessible at competitions. These advantages, however, are not enough to overcome the ingenuity of athletes with strong incentives to cheat.

We might imagine that people will not be motivated to violate restrictions on reproductive technologies, but this does not seem to be the case. A grief-stricken father phoned me six months after the suicide of his son and told me he wanted to clone the boy. The man was wealthy, determined, and would have done almost anything to accomplish this goal had the technology existed.

If we cannot keep performance-enhancing drugs out of sports, how can we effectively ban pharmaceutical and genetic enhancement in the rough-and-tumble of the real world? People will not agree that enhancement procedures are wrong and will resist intrusions on their privacy. Violators will be scattered and hard to identify, and punishment will be uncertain and slow.

It is hard to see what good will flow from blanket prohibitions today on the reproductive technologies of tomorrow. The landscape of their potential use is still so unclear that such laws will probably prove to be ineffectual for situations that are substantially different from what we now imagine. At best, legal sanctions will merely dull discussion of the coming challenges by creating the false impression that we can will them away. At worst they will stifle beneficial research, divide our society, and drive advanced GCT into the hands of the least responsible among us.

A Spiritual Crossroads

A key aspect of human nature is our ability to manipulate the world. We plant, we build, we dam, we hunt, we mine, and increasingly we do so on a huge scale. For as long as we have been able to, we have altered ourselves as well. We not only wear clothing, we pierce our bodies, tattoo our skin, cut our hair, and surgically sculpt ourselves. We add or remove hair, straighten our teeth, fix our noses, enhance or reduce our breasts, get rid of fat. We use drugs to reduce pain, lose weight, change moods, stay awake. The idea that we will long forgo better and more powerful ways of modifying ourselves is a denial of what the past tells us about who we are.

We are now reaching the point at which we may be able to transform ourselves into something "other." To turn away from germline selection and modification without even exploring them would be to deny our essential nature and perhaps our destiny. Ultimately, such a retreat might deaden the human spirit of exploration, taming and diminishing us. This seems particularly clear to the American psyche, influenced as it has been by the frontier. Many writers have described this exploratory exuberance. Early in the last century, the influential historian Frederick Jackson Turner put it this way:

> For a moment, at the frontier, the bonds of custom are broken and unrestraint is triumphant . . . Each frontier did indeed furnish a new field of opportunity, a gate of escape from the bondage of the past; and freshness, and confidence, and scorn of older society, impatience of its restraints and its ideas, and indifference to its lessons, have accompanied the frontier. What the Mediterranean Sea was to the Greeks, breaking the bond of custom, offering new experiences, calling out new institutions and activities, that, and more, the ever retreating frontier has been to the United States directly, and to the nations of Europe more remotely.

We often look to space as the next frontier — our expansion out into the solar system and beyond. Ultimately this may happen, but the next frontier is not outer space but ourselves. Exploring human biology and facing the truths we uncover in the process will be the most gripping adventure in all our history, and it has already begun. What emerges from this penetration into our inner space will change us all: those who stay home, those who oppose the endeavor, those tarrying at its rear, and those pushing ahead at its vanguard.

Albrecht Sippel, a German biologist who opposes human germline manipulation, had this to say when I suggested that there was no way to stop it: "If you suddenly would have an easy genetic way, let's say, to enhance intelligence or do better in school examinations, people would do it. They wouldn't wait fifty years to figure out whether it's really positive . . . I could say, as a European, let the Americans try. They are the guinea pigs. I don't want to be that guinea pig."

His is a reasonable position. The embrace of new technology is never uniform. Individuals and societies alike are comfortable with different levels of uncertainty and risk. But such preferences need not be cloaked in moral rectitude. After all, few people who philosophically oppose embryonic stem cell research would forgo the advances that arise from the enterprise, however morally inconsistent this position may be. If such research generates medical breakthroughs to treat Alzheimer's or heart disease, or slows some aspect of aging, we will want these benefits. The same is likely to be true of the possibilities of germinal choice.

Earlier, I discussed why outright bans on advanced germinal choice technology will eventually break down, but such bans are problematic even now. While they may deter those scientists who would otherwise mount the more visible and expensive early projects attempting to demonstrate these technologies in humans, they will neither halt nor slow the broader scientific progress needed for GCT to be viable in actual clinical settings. So bans will shorten the lead time between early demonstrations and eventual clinical rollouts. And this time is precious, because early demonstrations are what spark the media attention and debate that help us iron out the problems and prepare for the changes to come. The longer our lead time, the more measured and thoughtful we can be.

Early demonstrations are the heralds of what is coming, so we should be wary of stifling them. When an advanced GCT does not place a future child at inordinate risk, and a well-informed couple willingly chooses the procedure and pays for it, we should understand that however much we dislike what they are doing, society benefits. In essence, such couples are volunteering as the test pilots of human biological manipulation. These experiments will take place; the crucial question is whether society will find out what happens when they do.

This doesn't mean that we do not need to regulate the clinical use of these technologies. No one seriously thinks that widespread medical interventions on human embryos will occur without oversight. The issue is how close that oversight should be and whether its primary focus should be individual safety.

Policymakers sometimes mistakenly think that they have a choice about whether germinal technologies will come into being. They do not. If, in the mid-1700s, the British Parliament had banned the steam engine to try to stave off the industrial revolution, the action might have altered some of the details of the mechanization of human endeavor but would not have stopped it. The same is true of the computer and genomics revolutions of today.

At the heart of the coming possibilities of human enhancement lies the fundamental question of whether we are willing to trust in the future. Will we accept humanity's eventual transformation into something beyond human, or will we battle against it and try to protect those aspects of the human form and character that we see as intrinsic to our humanness?

I have argued that our exploration of these technologies is inevitable, but nothing can be truly certain. Critics who seek international bans on germinal choice technology seem to feel that they can head off the future I've described, or perhaps they merely believe that their struggle, though doomed, will at least preserve their own integrity. If some combination of these attitudes is commonplace, and I suspect it is, then the reproductive technologies that attend today's genomics revolution will provoke bitter cultural conflicts.

A belief expressed by many of those who would like to prevent the reworking of human biology and reproduction is, as I stated earlier, that we should not play God, although typically opponents express the idea in secular terms, such as our need to protect human dignity. Ironically, embracing the challenges and goals of these transformative technologies is an act of extraordinary faith. It embodies an acceptance of a human fate written both in our underlying nature and in the biology that constitutes us. We cannot know where self-directed evolution will take us, nor hope to control the process for very long. It will depend on our ongoing responses to continual changes that hinge not only on the character of future technologies we cannot yet glimpse but on the values of future humans we cannot hope to understand.

In offering ourselves as vessels for potential transformation into we know not what, we are submitting to the shaping hand of a process that dwarfs us individually. In secular terms, this is nothing special: we are merely accepting the possibilities of the advanced technologies we are creating. But from a spiritual perspective, the project of humanity's self-evolution is the ultimate embodiment of our science and ourselves as a cosmic instrument in our ongoing emergence. Rabbi Barry Freundel put it this way: "If G-d has built the capacity for gene redesign into nature, then He chose for it to be available to us, and our test remains whether we will use that power wisely or poorly."

One thing we can say with some confidence about the future we have embarked upon is that germinal choice technology may lead to considerable strangeness in the centuries ahead. Those who are happy to let GCT lead us where it may are trusting that our children, our children's children, and the many to be born after them will have the wisdom and clarity not to use this powerful knowledge in destructive ways. Those who fear that recent human history suggests that such trust is unfounded have no reason to think that we today can make better choices in this arena, much less project them forward in time. But if we can manage today's more pressing threats, coming generations will have their shot at tomorrow's. To accept a strange and uncertain future is not easy, because despite all our present weaknesses, at least we have an idea who we are. With technology so transforming the world around us, it might be comforting to know that the human family will, after all, long remain "human" and a "family." But perhaps this is asking too much — or too little.

One way to anticipate our responses to the challenging possibilities we are moving toward is to look at previous breakthroughs that have reshaped forever our vision of the world and ourselves. People sometimes snicker that the Catholic Church condemned Galileo. The Copernican universe is such a given today that it is hard to imagine a time when the Earth's place at the center of the universe was vital enough to Christianity's idea of heavenly order that the pope was willing to stake his infallibility on it. In the face of such assaults on a cherished worldview, denial is a natural refuge. On June 22, 1633, Galileo knelt before the tribunal that sentenced him and recanted his heresy:

> I must altogether abandon the false opinion that the Sun is the center of the world and immovable and that the Earth is not the center of the world and moves . . . With sincere heart and unfeigned faith I abjure, curse, and detest the aforesaid errors and heresies and generally every other error, heresy, and sect whatsoever contrary to the Holy Church, and I swear that in future I will never again say or assert, verbally or in writing, anything that might furnish occasion for a similar suspicion regarding me.

The next day his judges reduced his sentence from life in prison to permanent house arrest. He remained confined until his death, nine years later at the age of eighty-six, and his book *The Dialogue on the Two Chief World Systems* remained on the Index of Prohibited Books for more than two centuries. Not until 1992, more than 350 years later, did Pope John Paul II admit that the Vatican and Pope Urban VIII had erred.

The special significance of humanity seemed clear to Western thinkers in the Middle Ages: Earth was the center of the universe, and we were fashioned in God's image. The Copernican revolution shattered that notion, wrenching humanity from its exalted station and leaving it stranded on a peripheral planet circling one of many stars. The Darwinian revolution finished the job, leaving us fashioned not by divine consciousness but by random natural forces.

We survived these shocks and have even grown accustomed to our new place. Indeed, many take inspiration from the awesome immensity of the universe and ascribe a sacred quality to the natural environment and the evolutionary forces that molded us. There is majesty in these powerful forces and vast distances, resplendence in the eons of time that bore us.

We know all too well our limitations: our ineptitudes and weaknesses, our selfishness and egotism. No wonder the idea that we would attempt to fashion not only our future world but our future selves terrifies many people. In essence, such a process would replace the hand of an all-knowing and almighty Creator with our own clumsy fingers and instruments. It would trade the cautious pace of natural evolutionary change for the careless speed of high technology. We would be flying forward with no idea where we were going and no safety net to catch us.

Early in the year 2000, Bill Joy, the cofounder and chief scientist of Sun Microsystems, articulated this anxiety. In an article called "Why the World Doesn't Need Us," in which he urges us to relinquish genetic engineering and other advanced technologies, he writes, "If we could agree, as a species, what we wanted, where we were headed, and why, then we would make our future much less dangerous." To imagine that we could ever agree on such things, however, is the ultimate naiveté. Our future will emerge from the same chaotic jumble of trial and error and individual action and reaction that has formed our present. And this is fortunate, because we express a collective wisdom in our actions that far exceeds that of any commission of experts or council of elders who would guide us. Idealistic movements professing visions of a greater good have perpetrated some of the greatest evils of history. If instead of blinding ourselves with utopian images we admit that we don't know where we are headed, maybe we will work harder to ensure that the process itself serves us, and in the end that is what we must count on.

9

The Enhanced and the Unenhanced

> Gradually, the truth dawned on me: that Man had not remained one species, but had differentiated into two distinct animals: that my graceful children of the Upper-world were not the sole descendants of our generation, but that this bleached, obscene, nocturnal Thing, which had flashed before me, was also heir to all the ages.
>
> — H. G. Wells, *The Time Machine*, 1895

AS WE MOVE into an era of advanced germinal choice, children conceived with these technologies will necessarily intermingle with those with more haphazard beginnings. But how they will relate to one another in the long run is no more clear than whether a gulf will ultimately widen between them, partitioning humanity into the enhanced and the unenhanced.

The answers depend on which enhancements become feasible, their cost, who has access to them, who adopts them, and the nature of competing enhancements for adults. All this is as yet uncertain, but we can begin to discern some of the critical choices we will face. At so early a point in the development of GCT, identifying the policies that will serve us best is difficult, but spotting some that would serve us poorly is easy.

I have argued that germinal choice technology will offer us significant benefits and we will use the technology to acquire them. Moreover, the first wave of technologies offering substantive new human reproductive choices may be only a decade away. In-depth genetic testing, sophisticated preimplantation genetic diagnosis, egg banking, improved *in vitro* fertilization, and cloning are poised to transform our reproductive choices, while progress in genomics and with highly targeted pharmaceuticals will work in parallel by altering our perceptions of our genetic potentials, vulnerabilities, and handicaps. A decade or so beyond this first wave, a few rudimentary germline modifications may appear in special situations. And another decade beyond, more sophisticated and powerful germline manipulations may begin supplementing sophisticated genetic screenings and adult interventions. This timeline is little more than a guess, but whether substantive GCT enhancement arrives in ten years or fifty, the social and ethical challenges it brings will be similar.

A closer look at the possibility of overlapping effects of embryo screening procedures and early germline engineering suggests that preimplantation genetic diagnosis will likely be potent enough to provide significant human enhancement in addition to disease screening. Consider what would happen if parents wishing to enrich for a trait that is substantively shaped by genetics were to create a hundred healthy embryos, test them using PGD, and implant the one most predisposed toward that trait.

If such embryos could be selected, for example, for the gene variants responsible for a large portion of the genetic contribution to high IQ, the average score of children selected in this way might be nearly 120, well above the average score of 100 found in the general population and higher than nearly 9 out of 10 people. Moreover, this shift would take place in a single generation and use a proven medical procedure.

Such sophisticated embryo selection would be just as much a human enhancement as germline engineering. Indeed, no one would later be able to tell whether the lab had selected an embryo or modified one to obtain a particular genome. Here is another case where, if we continue to focus on the theological implications of laboratory procedures rather than on the results they bring, we will greatly weaken our attempt to deal with the approaching challenges.

To imagine that progress in germinal choice technology is irrelevant to the health of adults not planning high-tech parenthood is tempting, since the fateful meeting of sperm and egg that brought us into being is now beyond reach. Our lives, however, may be linked more directly to the arrival of advanced germinal choice than we might think. Adult enhancement, to the extent that it is feasible, rests on the same scientific foundations as embryo selection and will probably become available around the same time. The importance of nuclear transfer techniques to both reproductive cloning and regenerative medicine is no coincidence; such intersections will occur again and again.

Keeping in mind that genetic enhancement may be only a first step in humanity's coming journey of self-transformation adds valuable perspective to discussions of GCT. Fusion of human and machine in this century may be unlikely, but eventually the two may begin to join in important ways. If they do, the philosophical questions provoked by the coexistence of enhanced and unenhanced humans will arise once again. The future debate would not be about the displacement within our biological selves of the natural by the made, however, but about the displacement of the biological by the machine. This could make today's battles over mere biological enhancement seem quaint, because whatever these future humans may have become, they would have to grapple not merely with enhancement but with moving beyond biology itself.

The Enhanced

When bioethicists use the term "enhancement," they usually must confront the problem of defining normal human functioning, because they wish to differentiate between therapy and actual enhancement. The committee of the American Association for the Advancement of Science that considered germline policy in year 2000, for example, concluded that although "the use of IGM [inherited genetic modifications] to prevent and treat clear-cut diseases in future generations is ethically justifiable . . . IGM should be used only for cases which are clearly therapeutic." But such distinctions become arbitrary for such goals as retarding aging, which would be both an enhancement of our vitality and a therapy for age-related decline — in effect, a *therapeutic enhancement.*

I see nothing wrong with enhancement per se, so I use the term to mean any augmentation of attributes or overall functioning, whether or not it moves a person beyond our sense of normal human functioning. Abstract judgments about the value of particular characteristics and whether they relate to disease will tell us no more about how people will use germinal choice technology than they've told us about the use of cosmetic surgery or drugs. To see the future of GCT, we need to be more pragmatic and acknowledge that people want to be healthier, smarter, stronger, faster, more attractive. Enhancements are those modifications that people view as largely beneficial and that serve their goals. Virtually by definition, people seek such modifications.

A useful way of analyzing potential embryo enhancements is to categorize them by two measures: the degree to which the altered quality is health-related, and the magnitude of the embodied changes. Targeted traits will range from those that are clearly health-related — risks for heart disease, diabetes, or severe depression — to those that touch health less directly, such as obesity, and those that are largely cosmetic and idiosyncratic — hair color, musical talent, height, curiosity. We all would want our children to be at low risk for leukemia, but we might disagree about how tall or outgoing we'd like them to be.

We may best gauge the extent of a modification by comparing it to the typical range of human functioning in that realm. At one extreme are restorations of lost capacities — hearing for the deaf or improved immune response for those with compromised immune systems. In the middle are improvements that make people a little smarter, stronger, or taller, and that lift the underperformers to average level, and the average up to elite performance. At the other extreme are enhancements that carry a person beyond the normal human range, exceeding even today's elite performers — superhuman endurance, intellect, strength, or vitality.

Enhancements with different effects will present us with very different social, moral, and political questions. Many people, for example, have no problem with enhancements that are health related and that improve subnormal attributes — we generally call them therapies. Enhancements of idiosyncratic traits are more troubling because they can seem subjective and frivolous, though we know that those who seek cosmetic surgery often view it as critical to their mental health. But enhancements that would take us to elite or superhuman levels give most people cause for concern.

What is essential to realize about such interventions is that as the degree of enhancement rises, so will the technical difficulties involved. I cannot overemphasize this, because it is central to the future trajectory of human enhancement technology. Less extreme improvements will be much easier to accomplish and will even be available through embryo screening procedures, which do not manipulate genes. So initially GCT will offer less to those seeking superhuman performance than to those trying to avoid genetic impairments or improve some area of low or average performance. Widespread use of GCT would almost certainly raise average performance levels and improve health in coming generations, as well as narrow the spread between those with higher and lower potentials. This leveling does not arise from any imposed restriction on the technology. It flows directly from the step-by-step nature of technological advance, the greater complexity of more extreme enhancements, and people's tendency not to subject their children to unnecessary risks. Moreover, if adult enhancements become broadly available, they will lead to a similar flattening of the distribution of individual endowments.

Ridley Scott's 1982 film *Blade Runner,* like most science fiction, portrays genetic engineering as creating superhuman powers. The film's "replicants" are superior to mere mortals in most ways, but the tradeoff for their powers is extreme: after four years, they collapse and die. Roy, a replicant who returns to Earth looking for a way to escape his fate, crushes to death his creator, Tyrell, who tells him that it cannot be: "The light that burns twice as bright burns half as long. And you have burned so very very brightly, Roy . . . Revel in your time."

Such imagery disregards the immensity of the challenge of designing superhuman performance. Dog breeding illustrates the problem. Over thousands of years, emphasis on any one characteristic has brought tradeoffs with others. By amplifying specific wolf traits, we have bred specialist canines, not created superwolves. The saluki runs faster than any wolf. The bloodhound follows a scent better. The springer spaniel flushes game better. The toy poodle certainly is no superwolf. In their 1965 book *Genetics and the Social Behavior of the Dog*, John Paul Scott and John Fuller articulated it this way: "It is inconceivable that any particular domestic breed could compete with wolves under natural conditions . . . A wolf is a rugged and powerful animal adapted to life under a variety of adverse conditions. Consequently, no one of his behavioral capacities can be developed to a high degree . . . The idea that natural selection will produce a super-man or super-animal of any sort is an unobtainable myth."

We have seen too much progress in the intervening decades to be so sure that genetic engineering cannot create superior humans, but we are far from that goal. If the task were easy, natural evolution would have done it already. No such difficulty will keep us from improving average or below-average performance. All we have to do is copy nature. To give an embryo the genetics to achieve an adult height of eight feet without grave health problems would be an immense challenge; to achieve a height of six feet would not.

As GCT becomes increasingly potent, we will face tough personal decisions about what is best for our children and what risks and tradeoffs we will accept for them. There will be no simple answers. Our personalities and values will shape our attitudes. We will probably agree that certain types of manipulation are wrong, just as we agree that certain parental behaviors constitute abuse, and we will agree that certain enhancements exist that any responsible parent would make, just as we generally agree that kids should enhance their immunity by getting vaccinations. But there will also be passionate disagreements, and these cases will be very difficult to regulate.

Let's return to the example of the deaf parents who wish to have a deaf child. Germinal choice technology will make it possible, and as hard as it may be for someone with normal hearing to accept, preventing these parents from doing so would be dangerously close to coercive eugenics that targets the disabled. As long as deaf parents rely on embryo selection, stopping them from selecting an embryo destined to develop into a deaf child is tantamount to making them destroy that embryo.

The coming choices will force us to confront our attitudes about what constitutes a meaningful life, our responsibilities to others, our prejudices, and what we mean when we say that all potential lives are equal and deserve protection. In essence, we will soon have to face, in concrete human terms, the implications of our philosophies about human diversity.

Humans and Posthumans: Our Evolutionary Future

Humanity's manipulation of canine evolution has produced a wide range of breeds and served as an unwitting pilot project for our coming manipulation of our own evolution. In the early phases of human self-modification, the social constraints will be entirely different from those of canines, and the methods much more sophisticated, but scientists no doubt will encounter some of the same biological limits and possibilities.

Two critical questions that come up are whether this process will fragment humanity into independent breeds — future human Saint Bernard and dachshund analogs — and if so, whether they will persist and evolve into separate *posthuman* species.

First let's look at the idea of speciation. Despite the dramatically different shapes, sizes, and dispositions of dogs, *Canis familiaris* is still a single species. Reproductive isolation is central to speciation. Different species cannot interbreed under normal conditions. Such isolation is unlikely to occur in future human subpopulations. Not only will our offspring remain in close physical proximity, unless and until humans migrate out into the vast seas of space, but genomics and advanced reproductive technologies are breaching the barriers to genetic exchange even among different species. If scientists in Oregon can already give a jellyfish gene to a primate, surely we will continue to be able to exchange genes with one another.

In addition, species are biological forms that persist, averaging some four million years before extinction, according to the fossil record. If we succeed in progressively modifying our biology by altering our genes and supplementing our chromosomes, however, changes will be ongoing and new variants will emerge within a span of generations or centuries. Such posthumans could hardly be called "species."

Although even the concept of species may cease to be meaningful as reproduction shifts to the laboratory, the issue of whether the human community will eventually fragment into persistent independent groups remains. The only constant in a future of rapid biological manipulation would be evolutionary change itself. What could unite us in this future would be our common participation in this fluid, self-directed process rather than any transitory similarities in form. Seen in this light, strange as future humans may become if germline manipulation achieves its promise, they will still remain *human.*

In the past, the reproductive isolation needed to generate even the modest biological differences among human groups has required geographical or cultural separation. Both, however, are greatly diminishing because of increased individual mobility, modern communications, and softening cultural rigidities — trends likely to deepen despite strong opposition. Traditional Darwinian evolution now produces almost no change in humans and has little prospect of doing so in the foreseeable future. The human population is too large and entangled, and selective pressures are too localized and transitory.

In the future, however, the rapid technology-driven process of genetic design may achieve meaningful group-specific changes without reproductive isolation. With genetic refinements accumulating in the laboratory instead of in biological lineages, groups of individuals bound only by a common commitment to some specific enhancement could serve as a virtual test bed for refining genetic alterations. The spread of gene modules would not be by reproductive success but by reputation, word of mouth, even advertising. In essence, mimetic rather than biological mechanisms will drive the penetration of genes in the human population.

The underlying source of this profound shift in the evolutionary process is the external storage and manipulation of human genes. As the genetic constructs we provide our children are increasingly explored, maintained, and refined in laboratories, working their way into our hearts and our children's bodies by public relations and persuasion rather than sex, the cultural processes hitherto shaped by our biology will turn the tables and remold our biology.

The consequences of a similar externalization some five thousand years ago give a hint of the tremendous implications of the coming genetic breakthroughs. The development of writing allowed knowledge — which hitherto had been stored only biologically and passed imperfectly from one fragile brain to another — to be captured physically and copied as needed. The result was the accumulation, refinement, and spread of knowledge to an extent otherwise impossible. Civilization rests upon this.

Our genetics has been similarly constrained. Evolution, for all its awesome constructions, is ultimately a vast tale of trial and error — and a slow and cautious one for a large organism like us. Random change is more likely to be deleterious than beneficial, so modifications must prove themselves over many generations. But while nature has eons, you and I do not.

As researchers gather and correlate human genetic profiles and adult human attributes, they will be able to assemble and interpret information about the effects of various clusters of genes. When they identify favorable combinations, they will preserve them outside our bodies — in tissue cultures, in freezers, and on computers — and we will pass them on to our children if we choose to. We will make mistakes. But so do random variation and natural selection. The enormous collective project of conscious human evolution has begun.

The Tensions of Living Together

As we gain conscious control over our biology, we will transform the range of what is human by expanding our diversity. Whether this diversification of the human form and character will isolate us from one another and make us truly separate is uncertain. Here I refer not to physical isolation, though that might well occur, but to a separation of our spirits, our purposes, and our biological identities. Dogs and cats, for example, are distinct and live among us, but they are our pets. If groups of future humans come to see each other as different, will they be able to remain on an equal footing?

Such changes to whole populations will require widespread germinal choice, of course, but a gradual transition to laboratory conception will likely follow the arrival of comprehensive embryo screening and advanced IVF, as parents come to view this as protection for their children. Even those uninterested in enhancement may start to see it as reckless and primitive to conceive a child without prior genetic testing.

With the advent of germline engineering, however, human artificial chromosomes will probably render laboratory conception obligatory rather than optional. The union of egg and sperm from two individuals with different numbers of chromosomes or different sequences of genes on their extra chromosomes would be too unpredictable with intercourse. But laboratory conception may not be a burden because such parents will probably want the most up-to-date chromosome enhancements anyway.

This move from bedroom to laboratory conception is one that future humans are unlikely ever to reverse, because they will not want to discard the benefits residing on their artificial chromosomes. This change seems dramatic, but it is not as big a leap as it might initially seem. Laboratory conception is just one more step down the path we took long ago when our distant ancestors embraced fire, clothing, and other early technologies, beginning a cultural process that has continually deepened our connection with and dependence on technology.

In the future, laboratory-mediated conception may seem no more foreign than medically assisted birth does today. In 1900, few thought of giving birth in a hospital as "natural"; only 5 percent of births took place there. Today, in the United States, almost all births do, and some 30 percent are by cesarean section, a frequently avoidable procedure that is nonetheless readily accepted.

* * *

How germinal choice technology affects our future will hinge on who has access to it as well as on what it offers. If the technology is available to large numbers of people, it is unlikely to give rise to a narrow elite.

Regulatory and health-care policies will be important factors in determining how broadly available GCT becomes, but the nature of the technology itself may play a more critical role. Whether GCT is a free healthcare benefit or is for sale on the open market, the more complex and individualized the technology, the more expensive and less widely available it will be. With healthcare plans the costly procedures are rationed; on the open market, only the affluent can afford them.

Different technological approaches will likely lead to procedures with different costs. To alter specific genes in place in an embryo's genome, for example, would be expensive because it would require a customized research effort for each embryo. Artificial chromosomes, on the other hand, might allow cheap enhancement for the many, because robotic devices could load them with a tailored package of off-the-shelf gene modules, and validate and test them before injecting them into embryos. So government policies that encourage research to refine artificial chromosomes and other technology platforms suited to widely available GCT might push development in this direction.

An altogether different strategy is to focus on access to the technology by attempting to control its clinical use. Such an approach poses a significant risk because it leads to categorical bans that, as previously discussed, will reserve the technology for narrow segments of society. Provision of free universal access to major aspects of GCT would align better with our ideals of equal opportunity for children and might be surprisingly affordable. If the price of a full GCT procedure could be kept down to, say, $6,000 a baby, this would be roughly equal to the average yearly expenditure on a student in public school in the United States.

As GCT begins to offer parents truly meaningful possibilities, our regulatory policies will have significant consequences for society. The first important choice we face will be our handling of advanced PGD, IVF, and egg banking. If tests to screen for almost all genetic diseases, for example, become available, but primarily to the affluent, such disorders will turn into diseases of the disadvantaged. Our policies will become even more crucial when we can screen embryos for genetic potentials.

As society moves closer to becoming a meritocracy, the most talented from all ethnicities and backgrounds will intermingle, form partnerships, and mate with similarly talented and successful others. Over time, this self-sorting will tend to divide society, increasingly distancing the more gifted from the less. Narrowly limited genetic screening and enhancement technology would accelerate such divisions and reinforce privilege, whereas broadly available technology would counteract them.

Not long ago, restricting access to education was one way of reinforcing class divisions, but we work so hard now to provide every child with the education to reach his or her potential that this repartitioning of society by talent and intellect, rather than by family and status, is already well under way in many countries. In the United States, the student bodies at elite institutions are ethnically and culturally more diverse than ever, but they are drawn from a narrow segment of the population. In 1990, Yale and Harvard together enrolled 1 in 400 of all freshmen at four-year colleges, but that included 1 in 10 of the small number of students scoring above 700 on the verbal portion of the Scholastic Aptitude Test, or SAT. This aggregation of an intellectual elite at the top universities is a new phenomenon. In 1950, such schools were ten times less selective when it came to standardized tests.

Kids soon learn how competitive the world is and where their talents do and do not lie. We have all been through this. If we were astute, lucky, or found good mentors, we ended up doing what we were best at. Some of our aptitudes emerged from our experiences; others — those innate talents that come so naturally we may take them for granted — came straight from our biology. People without the special talents and attributes that our society values — those who are clumsy, inarticulate, unattractive, slow-witted; those who would find it wonderful just to be average — are at a great disadvantage. Their hopes and aspirations may have always matched their lesser potentials, but more likely their dreams had to shrink one disappointment at a time.

Perhaps a mother who is unattractive remembers what it was like to suffer the teasing of her classmates and recalls her struggle for acceptance. Perhaps a father who was short and weak recalls being picked on as a boy. Perhaps a young man remembers watching others easily answer questions he could not fathom and thinks back on how humiliating it was to be dropped to the "slow" group. Maybe a young woman wanted to be a writer, but could never bring any magic to the words she wrote. These wounds heal, but they do not go away. Saying that we all have special, different talents and need to find them is too glib. Think about that person who was not bright or athletic or musically gifted, who felt lucky just to get by.

We have no choice, of course, but to play the hand we are dealt. But at the same time, we strive to protect our children and give them the breaks we never had: the education we couldn't afford, the family stability we wanted, the wealth we dreamed of, the guidance we needed. Society applauds these efforts but will be wary of parents who try to help their children through genetic interventions. Safety aside, though, why shouldn't we try to give our future children the talents we did not have or eliminate deficiencies that held us back? If we could make our baby smarter, more attractive, a better athlete or musician, or keep him or her from being overweight, why wouldn't we?

One social problem that might attend germinal choice technology, if it really can give our children raw talents, would be that such enhanced abilities would soon be less special. As in Garrison Keillor's Lake Wobegon, all the children would be above average. To the extent that talent and good health are heritable, children of some parents have an edge. Show me the brilliant intellectual who does not expect his child to be near the top of the class, the sports superstar who does not expect his child to have athletic gifts. Such kids may not turn out to be the smartest or the most talented, but they will probably do fairly well. Their genetics is not the whole story, but it is important. There is a reason adopted children tend to resemble their biological parents more than their adoptive ones: life does not start from scratch each generation; it takes from the past.

With the completion of the sequencing of the human genome, it has become fashionable to make a point of saying that we differ from one another in only 1 in 1,000 of our DNA bases. We are 99.9 percent the same as our fellow humans, whoever they may be. This statement is reassuring and politically correct, but misleading. We only have to look around us to see the extraordinary differences among us. Biological diversity is real. We come in a multitude of shapes and sizes. We have distinct personalities and temperaments. We possess various talents and vulnerabilities. We draw much of this from our genetic constitutions.

How can this be when our genetics are 99.9 percent the same? We see the answer when we realize that our DNA sequence is about 98.5 percent the same as a chimpanzee's, perhaps 85 percent the same as a mouse's. Open up a mouse and you find a heart, lungs, intestines, bones, nerves, muscles. Mice are close cousins to us. And when it comes simply to having homologous genes rather than exact DNA sequences, the similarity between all life is even clearer. Some 98 percent of the mouse's genes are ours too, 60 percent of the fruit fly's, and more than 25 percent of those of a banana. All life has cells. These cells divide in the same ways. They regulate their DNA and manage their metabolism and cellular communication in the same ways. They have the same basic biochemistry. Our genetic similarities come from the fundamentals we all share. Of course you and I are nearly the same. We are both animals, both vertebrates, both primates — both humans. The differences between us are subtle, but that doesn't mean we shouldn't care about them. A difference of 1 in 1,000 bases between any two people is not trivial, even though it is much less than the difference between, say, two chimpanzees. It amounts to 3 million differing bases in their individual genomes. Sure, the vast majority will be scattered through the so-called junk DNA between their genes, and of the 150,000 or so differences in their actual genes, most will be neutral and not lead to any functional differences. But a single base can be the difference between vibrant health and early death. Parkinson's disease comes from a single changed base. So do sickle cell anemia and hemophilia. A single base difference can make the fingers on a person's hands resemble toes, or cause mental retardation.

We do not yet know what percentage of two people's genes differ in meaningful ways, but a good guess is around 10 to 20 percent — several thousand genes. Moreover, because our major competitors for just about everything in life are other people, we are fine-tuned by evolution to be highly sensitive to the minute distinctions among us. We don't care that we are all mammals, all primates. These are givens. We care about our differences. All people might look pretty much the same to a space alien or a mosquito — or even an evolutionary biologist — but not to a coach trying to build a winning sports team or to someone looking for a mate.

As we untangle our genes and learn to select and alter them, some parents will want to give their children endowments they themselves could only dream of. If such interventions become commonplace, the result will be revolutionary, because it will be a major step toward equalizing life's possibilities. The gifted of today ultimately may not welcome such a leveling, because it would diminish the edge their children enjoy and make society very competitive, even for the best endowed.

If GCT enhancements prove feasible, eventually the mass of humanity will seize the power to enrich its children's natural endowments. Strong voices will oppose this, but most of the warnings about the danger of eugenics and the threat of lab-built humans will come from people with the most to lose — the well-endowed elite. Surely theirs are the children who would ultimately suffer from the arrival of a genetic bazaar where all parents can obtain equivalent talents and potentials for their children.

Today's intellectual elite might not want to live in a world as aggressive, competitive, and uncontrolled as the one that would emerge from universal access to potent germinal choice technologies, so their distaste for the technology may deepen once its true implications become clear. Such resistance would be reminiscent of earlier elites who liked society the way it was and tried to protect their privileged position. The wellborn of an earlier time were right to fear the political reforms that broke down class divisions. Now a new elite may wince, because if the God-given gifts of talent and intelligence that have raised them above the throng are suddenly laid out for everyone else, their future would not be so secure.

The concern about preserving the ideals of an egalitarian society and preventing the creation of a genetic elite that inspires many of the philosophical objections to GCT among intellectuals is ironic, both because they themselves are among the elite and because any enduring effort to block these technologies will restrict access to them, compounding rather than alleviating genetic advantages.

To look at the possible implications of advanced germinal choice technology more concretely, let's imagine a best-case scenario: science and policy combine to make germinal enhancement widely available, relatively commonplace, and largely under the control of individual parents.

In such a society, many parents might shun GCT, but others would embrace it enthusiastically. With time, people's genetics would become a manifestation of their parents' values and predilections. Initially, the differences between the enhanced and the unenhanced would be only statistical, in that those with enhancements would tend to outperform many, but not all, of those without enhancements. But as the technology grew more potent, less overlap would exist between the two populations, and as this became clear to parents, many of the children of those who had shunned the technology would likely enhance their own children, to keep them from being at a disadvantage.

A similar story would be played out globally, as countries that initially blocked GCT gradually felt compelled to amend or repeal their laws and accept it. Access to advanced technology typically flows from national wealth, but adoption of GCT may hinge more on the religious and cultural traditions in particular regions. Some of the richest nations could easily be the most resistant to the technology. Eventually, however, they probably will have no choice. What, after all, would a country that bans advanced embryo diagnosis do if other nations were to embark on popularly supported eugenic programs aimed at dramatically raising the average IQ of their next generation?

Breeds Apart

As we move from embryo selection to direct germline enhancement, one might imagine that devising artificial chromosomes that can enhance one embryo as effectively as another would ensure that humanity would not split into separate breeds, since future parents, whatever their own particular genetic endowments, would be able to select their children's genetic modules from an expanding common library of enhancements.

Enabling couples to give their future children genetic predispositions differing from their own, however, does not necessarily mean that they would. Our experiences, associations, and natures circumscribe our values and attitudes. Parents with general competencies might well consider such balance good for their children, while parents with narrow, highly specialized talents might see greater specialization as preferable.

As we increasingly manifest our aptitudes, temperaments, and philosophies in our children through our decisions about their genetic makeup, a self-reinforcing channeling of human lineages will likely develop. Family names once denoted family professions that persisted for generations. The Cooks, Fishers, Smiths, Taylors, and Bakers of the world could no doubt uncover the corresponding trades among their ancestors. Perhaps in the future, clusters of genetic predispositions, lifestyles, and philosophical orientations will arise that are equally persistent. Families of musicians, politicians, therapists, scientists, and athletes would not be locked in by social constraints and limited opportunities, however, but by tight feedback between genetic selection and the values, philosophies, and choices that both author that selection and flow from it. Such future human specialization might be far deeper than that which has occurred historically. Maintaining it would require an ongoing cycle of renewed choices that couples, at least theoretically, could break when they decided to have children, but most probably would not.

Ideally, the resulting partitioning of the social landscape would proceed according to individual predispositions and desires rather than some preexisting template imposed on unwilling populations. But the possibilities of abuse by governments or individual parents who breed children for their own purposes should rightly give us pause. We must remember, though, that tyranny and child abuse require no advanced technology, and whether either would be changed much by the presence of germinal choice is highly debatable.

Even disregarding outright abuse, scenarios of human design are jarring, if not frightening, because they evoke troubling images of freakish human forms. While we should not dismiss such images entirely, neither should we allow them to grow in our minds to the point where they oversimplify and distort a future landscape whose complicated topography is not yet defined. Rare, special attributes such as photographic memory or extraordinary athletic ability may become both more extreme and more commonplace, but that does not mean they will be grotesque. As for the fear that parental choices might become too uniform, children would be as unique as they are now. The multiplicity of individual experience molds an infinity of expression from our genes, even if they have been chosen.

Despite the occasional exception, as in the case of parents who select deafness, when we are able to choose our offspring's genetic predispositions, we will probably opt to avoid most of the genetic disabilities and vulnerabilities that afflict us today. In this limited sense, early applications of preimplantation genetic diagnosis will narrow human diversity, but the polio vaccine did as much and brought few complaints. And as sophisticated embryo screening and germline manipulation begin to enrich enhancement possibilities, no doubt clusters of attributes will be reinforced, which in time will expand diversity.

Today, when that rare combination of genetic and environmental factors comes together to create genius in one of its many guises, the combination usually disappears in the next generation. This happens because contributing environmental influences do not recur and because genetic constellations dissipate during genetic recombination. In the future, parents not only might preserve key aspects of such genetic influences through embryo selection, they might also have the knowledge and the means to push those talents even further, by creating environments that reinforce them and by refining chromosomal additions.

What aspects of themselves people will want to boost or moderate is hard to say. But taken together, their choices will have a powerful effect on society. Children's biological predispositions will come to reflect parental philosophies and attitudes, and thus children will manifest the ethos and values that influence their parents. Consider gender. Many couples would make different choices about the attributes of boys and girls. Thus, GCT might translate cultural attitudes about gender into the biology of children. If a society believes that women are (or should be) more empathetic and supportive, and boys more aggressive and independent, then whether or not these gender specificities are true now is not as important as the likelihood that they will gradually become true.

Because our notions of personal identity are specific to particular cultures and times, purely cultural distinctions could become more embedded in our genetics and may increase the biological differences among human populations. Each culture assigns its own value to traits such as calmness, obedience, and curiosity. To the extent that genes can influence these differences, GCT might reinforce them.

Many social commentators today complain about the homogenization of culture, brought on by global commerce and communications. The arrival of advanced, widespread germinal choice technology may counteract that trend by allowing people to infuse some of their cultural differences into the biology of their children. In a mixed cultural environment like that in America, of course, these effects would be played out on a national stage. Current debates about whether some of the differences among ethnic and racial groups are cultural or biological will soon become irrelevant, given the coming interdependence of the two. In any event, once we can fashion our children's biological predispositions, many cultural and personal influences will feed directly into biology.

Enhancement will be not a single dimension of change but a wide range of modification and augmentation, superimposed on the broad distribution of naturally occurring human qualities, so distinguishing between the enhanced and the unenhanced will initially be difficult, if not impossible. But germinal choice will eventually become so commonplace that the question won't even be interesting, especially given that potent quasi-medical pharmaceutical enhancements for adults will probably also become widespread. If so conservative a group as the Amish are willing to seek gene therapy, virtually no one will forgo drug, technological, and genetic enhancement once it is safe, reliable, beneficial, and perhaps even fashionable.

In the future, humanity will be an ever-shifting mélange of those who are biologically unaltered, those with improved health and longevity, and those with sundry other enhancements. In essence, we and our children increasingly will be reflections of our personal philosophies and values. Where today we sculpt our minds and bodies using exercise, drugs, and surgery, tomorrow we will also use the tools that biotechnology provides.

We cannot say what powers future humans will assume, what forms they will take, or even if they will be strictly biological, but we can be certain of one predisposition they will have. They will be committed to the process of human enhancement and self-directed evolution. This we know, because without this commitment they would lag behind and be displaced by those who are more aggressive in this regard.

But the immediate cultural landscape wrought by germinal choice and biological manipulation will be more familiar than we might think. We are used to enormous human diversity. An anorexic eighty-pound fashion model, a four-hundred-pound sumo wrestler, a tiny svelte gymnast, and a towering basketball center are very different from one another. So may be their lives and passions — or those of a deaf-mute and a concert pianist, for that matter. We also are used to enormous technological gulfs between people: the vision and hearing of those with and without televisions and telephones differ greatly.

Whether our differences today are primarily the result of genetics, culture, technology, or education, they are real and they permeate our lives. Many of us revel in the giant and diverse human aggregations that are our modern cities; others simply cope with them. The enhancements brought by germinal choice will not soon sweep us into a realm so alien that we could detect the change on a stroll through a crowd but the changes will affect us profoundly.

The biggest challenge will be our changing image of ourselves. At the outset I said that these new technologies would force us to examine the very question of what it means to be human. As we follow the path that germline choice offers, we are likely to find that being human has little to do with the particular physical and mental characteristics we now use to define ourselves, and even less to do with the methods of conception and birth that are now so familiar. Adjusting to new possibilities in these areas will be hard for many of us, because it will demand a level of tolerance and acceptance that until now has been the exception rather than the rule. But perhaps the drama of the shift will itself ease the change by capturing our attention and forcing us into new patterns of relationship. Until now, to accept each other we often have had to pretend that we are all the same, but maybe when we see that we are all different and unequal — increasingly so — we will learn to accept our differences.

As we move into the centuries ahead, our strongest bond with one another may be that we share a common biological origin and are part of a common process of self-directed emergence into an unknowable future. Seen in this light, the present differences among us are trivial, because we are companions in transition and are likely to remain so.

Perhaps this state of transition is what has always defined us. The mechanisms of change were different in previous eras, but the culturally and technologically driven process of becoming that which we are not, of changing the world around us and our own selves, is not new — only the pace and depth of change are truly unprecedented.

A World Aborning

To some, the coming of human-directed change is unnatural because it differs so much from any previous change, but this distinction between the natural and the unnatural is an illusion. We are as natural a part of the world as anything else is, and so is the technology we create. As we consciously transform ourselves, we will become no less human than we became tens of thousands of years ago when we embarked upon a course of self-domestication and began, quite unconsciously, to self-select for the human qualities that enable us to live and work together effectively. That we are uneasy about what lies ahead is not surprising. The arrival of GCT signals a diffuse and unplanned project to redesign ourselves. But it is neither an invasion of the inhuman, threatening that which is human within us, nor a transcendence of our human limits. Remaking ourselves is the ultimate expression and realization of our humanity. We would be foolish to believe that this future is without peril and filled only with benefits, that these powerful technologies will not require wisdom to handle well, or that great loss will not accompany the changes ahead. We are beginning an extraordinary adventure that we cannot avoid, because, judging from our past, whether we like it or not this *is* the human destiny.

It brings to mind the advertisement that Sir Ernest Shackleton, the renowned British explorer, placed in 1912 when he was recruiting a team for an expedition to cross the icy Antarctic continent on foot: "Men wanted for hazardous journey. Small wages. Bitter cold. Long months of complete darkness. Safe return doubtful. Honour and recognition in case of success."

Some five thousand people responded to his call. Shackleton selected twenty-seven and began the journey. Their ship became frozen in place just off the Antarctic coast and was later crushed by the ice, but after a harrowing and nearly unbelievable two-year ordeal, they all returned safely. Shackleton looked back on it in his diary: "In memories we were rich. We had pierced the veneer of outside things. We had suffered, starved and triumphed, groveled down yet grasped at glory, grown bigger in the bigness of the whole.

We had seen God in His splendors, heard the text that nature renders. We had reached the naked soul of man."

Our journey into our own biology is very different. The endeavor is collective rather than individual, its course encompasses centuries rather than years, there will be no return, and the voyage is as spiritual as it is physical. But we too are entering uncharted territory. We too will no doubt face adversity. And the destination may prove less important than the journey itself. As we pierce the veneer of inside things, we too may reach the naked soul of man.

We have created artificial intelligence from the inert sand at our feet through the silicon revolution, we are moving out into space from the thin planetary patina that hitherto has held all life, we are reworking the surface of our planet and shaping it to suit us. These developments will transform the world we inhabit. Amid all this, could we really imagine that we ourselves would somehow remain unchanged? Or that we would want to? If we were to succeed in turning back from the path of self-modification now opening before us, we would not be pleased with the result, because ultimately we would find ourselves in a world so different from the one that spawned us that we would feel estranged and adrift. Adaptable as we are, to remain at home in the world we are forming, we will have to adjust ourselves to cope with it.

At the end of the nineteenth century, visionary biologists imagined a bleak human future. Our very successes, by softening natural selection and saving those who would otherwise die young, seemed to ensure that the human species would slowly deteriorate. Germinal choice technology frees us from this fate, but it brings other, more immediate threats.

As we enter into advanced reproductive technology, we would do well to recall the Nazi concentration camps. Eugenics, as practiced in the first half of the past century, attests to the horrors of governmental abuse, and although Germany was the most egregious case, it was not alone. In the 1920s, the eugenics movement, which was often called "race hygiene," was centered in the United States and Great Britain, and included adherents in Poland, France, Italy, Scandinavia, Japan, and Latin America. Many of these programs were voluntary, but some were not.

Some say that if we do not learn from history, we risk repeating it, but the challenge is always to understand what history is telling us. The lessons of past eugenic abuse do not concern technology, biology, or human reproduction, but nationalism, totalitarian regimes, individual freedom, and tyranny. Government abuse is what we must fear, not germinal choice technology. GCT is not a weapon, and the chaos of countless individual genetic choices by individual parents is not a threat, especially if the choices are circumscribed by modest oversight. Some push for uniform global policies, but these raise specters of the same governmental abuses that history warns us about. Far better that we find our way in this coming journey by trial and error — cautious, informed trial, of course, and as little error as possible — but trial and error nonetheless.

There is no way we can permanently forgo these enhancement technologies if they prove robust and useful. Those who would shun healthier constitutions and extended lifespans might hope to remain the way they are, linked to a human past they cherish. But future generations will not want to remain "natural" if that means living at the whim of advanced creatures to whom they would be little more than intriguing relics from an abandoned human past.

What is occurring now is no less than a birth. The occasion may prove messy and painful, but it carries the wonder of new life and new possibilities, the promise of growth and achievement. Humanity has been building toward this moment for tens of thousands of years. Conception took place long ago when we first chipped stone tools and used fire. Quickening came with early agriculture, writing, and the formation of larger human communities. Now the contractions are forceful and rapid. The head is beginning to show. Will we suddenly lose our nerve because of the realization that life will change forever and because we can barely guess the character of this child of our creation? I hope not. We cannot push the head back, and we risk doing ourselves grievous harm if we make the attempt. We may not like the future we are creating, so vastly will it ultimately differ from our present. Yet our descendants — those beings who are the product of today's crude beginnings at unraveling our biology — will be unable to imagine living without the many enhancements that we will make possible for them.

A thousand years hence, these future humans — whoever or whatever they may be — will look back on our era as a challenging, difficult, traumatic moment. They will likely see it as a strange and primitive time when people lived only seventy or eighty years, died of awful diseases, and conceived their children outside a laboratory by a random, unpredictable meeting of sperm and egg. But they will also see our era as the fertile, extraordinary epoch that laid the foundation of their lives and their society. The cornerstone will almost certainly be the reworking of human biology and reproduction. To me, being here, not only to witness but to participate in this unprecedented development, is an amazing privilege. But we are so much more than observers and architects of these changes; we are also their objects.

Public policies about germinal choice technology will be effective only to the extent that they are prescient enough to elicit technology that succeeds in promoting future research and development. Whether DNA chips, advanced *in vitro* fertilization, and human artificial chromosomes will provide the foundation for germinal choice or whether some other cluster of technologies will fill that role remains to be seen. In either case, we will work toward embryo screening and other germline procedures that are either cheap and accessible to everyone or expensive and accessible to only a few.

If there is a window of opportunity for government to influence the future path of these technologies, it is unlikely to last for long. Only a few countries now have the capacity to realize them. If these nations move toward workable GCT and responsible strategies for using the resulting reproductive procedures, they may shape the basic approaches that become dominant. If they restrict the development of GCT and simply continue probing the fundamentals of our biology, unknown others will take the critical early steps and determine the shape of GCT for the immediate future.

The great challenge is not how we handle cloning, embryo selection, germline engineering, genetic testing, genetically altered foods, or any other specific technology. We will muddle through as we always have. Unlike nuclear weapons, these technologies will be forgiving; they carry no threat of imminent destruction to multitudes of innocent bystanders. The crucial question is whether we will continue to embrace the possibilities of our biological future or pull back and relinquish their development to braver souls in more adventurous nations of the world.

Many European countries have already made a provisional decision in this regard. In part because of their sensitivity over the eugenic abuses of the past, they are forgoing these technologies and trusting others to forge them. I suspect that this stance on so central an element of our future will be temporary, but in any event, Europe will have a decade to mull over the matter before GCT emerges in a serious way.

As I see it, the coming opportunities in germinal choice technology far outweigh the risks. What is more, a free-market environment with real individual choice, modest oversight, and robust mechanisms to learn quickly from mistakes is the best way both to protect us from potential abuses and to channel resources toward the goals we value.

Appendixes

Acknowledgments

Notes

Bibliography

Index

Appendix 1 Regulatory Paths in the Era of Germinal Choice

In the main text, I proposed a relatively hands-off approach to regulation, while maintaining that eventually we will require sufficient oversight to avoid egregious misuses of the additional power that parents will be able to exert over their children. Here I extend these arguments to provide an overview of the challenges that will face regulators as germinal choice technology moves forward. Looking at specific interventions now is premature, since they are still so uncertain. But it is useful to consider why regulation will be so difficult, and what competing interests we will need to balance in managing the clinical use of these technologies.

Three overlapping domains of interest exist in the arena of germinal choice:

- The interests of society at large — our need to maintain the integrity of various governmental, religious, and social institutions, as well as preserve the shared traditions and values that underlie society.
- The interests of parents — their desire to have children, to make choices based on their own values, and to protect their families and help them thrive.
- The interests of future children — their needs for sustenance, love, acceptance, care, and protection from both the state and, on occasion, their parents.

When interests in these realms align, we have no problem. But often they do not, in which case our regulatory environment must mitigate and resolve the conflicts. Regulations in reproductive matters can have strong biases toward any of these three constituencies, and each bias carries its own set of problems.

A bias toward societal interests, by its very nature, can bring the most widespread abuses of individual rights. The one-child-per-family policy in China is one instance of such social engineering. The bans on birth control in various Catholic countries in Latin America show a different facet of interventions: the effort to legislate morality. Such policies run roughshod over private and individual interests. It is here that most eugenic policies sprout, with their top-down approach to shaping the population as a whole. Whether such programs intend to improve a population's genes, reduce the number of disabled who might burden public resources, or alter reproductive patterns in various socioeconomic or ethnic groups, they are all essentially efforts to influence the constitution of the next generation in a supposedly desirable way by controlling individual reproductive behavior. Such policies come in many forms, because the political and moral philosophies that shape them can be so different. As the potency of GCT increases, this realm will need to be watched vigilantly, because it can be fertile soil for grand social experiments. Our newfound powers in human reproduction will make such efforts tempting, but we have the lessons of history to remind us how easily even the most innocuous family-planning programs can mutate into malignant forms.

Parental freedom has tended to be the dominant force in procreation through most of the developed world. Women now are relatively free to have children when and as often as they wish, and, subject to some restrictions, they can also terminate pregnancies or avoid them. Indeed, in the United States, the courts now interpret procreation as a basic constitutional right. Efforts have been made to restrict parental discretion in some areas, but these attempts have met strong resistance. The most contentious issue is, of course, abortion, but there have also been proposals to restrict childbearing in women who cannot care for their kids, who abuse drugs, or who might otherwise damage their children-to-be. Nonetheless, parental freedoms remain largely intact.

Finally, the interests of the child-to-be deserve consideration. Without this concern, the regulatory issues surrounding reproduction would just be another facet of the long-standing tension between individual and societal interests. Parental use of GCT will influence future children in ways that are difficult if not impossible to gauge in advance. Critics of human cloning in the United States, for example, frequently assert that being a clone would damage a child psychologically, and that therefore the practice, even if it were medically safe, should be illegal. This argument is unconvincing, given that identical twins (the only clones we have information about) do quite well. Other potential threats, however, are less easily dismissed. What about a child selected by religious fundamentalists to be obedient, a beautiful girl selected by parents hoping to turn her into a child star, or even an extremely gifted child chosen by parents with no such gifts?

In the United States and many other countries, broad parental latitude about conception and pregnancy has been the basis for most public policy on reproductive technologies. As long as parents can exert relatively limited control over the natures of their future children, this works well, but increasingly the boundaries of parental rights will have to be examined. Not only do the emerging technologies seem different from "natural" reproduction, the possibilities they are creating — from *in vitro* fertilization procedures that create large "litters" that parents cannot support, to the engineering of problematic temperaments in kids — will bring up real conflicts between parental desires and the possible interests of children and society as a whole.

Where we rely on religious and governmental perspectives in our design of parental constraints, we will be entering risky territory because visions of the common good are so subjective and so easily manipulated in self-serving ways. The welfare of future children is a different matter. Clashes arising between the desires of prospective parents and the interests of their children will likely bring the most legitimate and thorny problems for future regulation, and the area warrants caution.

Here are a few suggestions for minimizing the conflicts to come:

- We should deal with actual rather than imagined problems. If we write preemptive regulations based on vague fears rather than real occurrences, many benefits will pass us by. Coming up with utopian and dystopian scenarios is easy, but they tell us more about their author's own hopes and fears than about what the future will bring.
- We need to be vigilant about egregious cases of self-serving parents who attempt to bear children to further their own interests — the creation of passive servants or elite performers to make money — but this will be hard to untangle from normal parental motivations except in extreme cases. To put much faith in questionable projections of the imagined happiness of hypothetical children will not provide much useful guidance.

- Although it may not be achievable in the current climate in the United States, future children would be well served by separating the issue of their protection from the debate about the legal status of embryos. If we can guard the interests and dignity of the real living children of the future, we will have accomplished something extraordinary. Attempts to convert the complex challenges of GCT into a simple question of embryo survival lead nowhere. This is a theological battle of the present. That such issues drive policy in religious states is entirely appropriate, but not in multicultural secular democracies. *In vitro* fertilization is life-giving, but leads to the destruction of embryos because not all those that are created are used. Germinal choice technology will be the same. Focusing on children rather than embryos would have major practical implications. For example, consider the debate about gender selection, which some feminists see as discrimination against the unborn. If the treatment and dignity of the born are put first, then as long as parental choices about gender are not so unbalanced as to skew the population, the procedure will not be a problem.
- We should think more about outcomes than about technology, and try to broaden our regulations and apply them to all conception. In other words, consider safety. If a couple considering germline engineering had a 1-in-5 chance of producing a retarded child, the procedure would be too dangerous for most parents and physicians to tolerate. But if we decide to outlaw it, we should think more carefully about how to handle the case of a woman who knows she has Huntington's disease. Her child would have a 1-in-2 chance of suffering its dementia and fatal decline. To be consistent, if the woman disapproved of abortion and refused to be genetically screened for the disease, we would have to prohibit her from having a child. If we prefer to focus on the process rather than the outcome, we would do well to decide what makes "natural" reproduction sacred in a way that changes our regard for risks to a potential child. Tay-Sachs disease, which brings a painful early death, highlights this. If two parents are carriers and conceive a child, the infant has a 1-in-4 chance of suffering this fate — higher than the risk we assumed for germline manipulation. But are we willing to consider prohibiting women from having children without genetic screening? As GCT improves, these questions will become even harder. If a parent could use sophisticated fetal screening to eliminate 90 percent of the potential

> problems from our hypothetical germline manipulation, only a 2 percent chance of retardation would remain. If this seems too high to allow, then we might also consider whether we should prohibit women at the age of forty-five from having a child if they do not intend to screen for Down syndrome, since they face a 3 percent chance of bearing such a child.

As laboratory-mediated reproduction becomes more commonplace, it will be ironic, though not surprising, if we develop standards for GCT that make it much safer than natural conception and yet for the sake of tradition tolerate high risks in some situations of natural conception. But once we can measure the risks, I suspect that we will not entirely ignore them, especially if society rather than individual parents bears the cost of caring for disabled children.

A swirl of competing interests is bound to impinge upon the emerging possibilities of GCT. Moreover, our decisions about these technologies are unlikely to remain solely in the realm of germinal choice; they will seep into our handling of natural procreation as well. Any consensus on these questions is doubtful, even without bringing religion into the picture. We simply will have to struggle to balance the interests of parents, children, and society. If we are flexible and open-minded, we may be able to learn quickly from our mistakes.

The challenges facing us do not pose a choice between right and wrong, but a need to mediate between right and right. Given the rapidly changing technological landscape, the uncertainties about the nature of the possibilities before us, the ill-defined and changing risks of various procedures, and the as yet unknown ways that people will try to use these emerging technologies, we need to retain the capacity for modifying our approaches. This suggests that we should avoid sweeping legislative responses that are difficult to amend, rely as much as possible on narrow administrative guidelines that can be adjusted in response to new information, and use case law to deal with the individual conflicts that arise from early implementation of GCT. Only when a real and pressing need exists should we resort to more extreme legal and legislative tools.

Appendix 2 Challenges to Come

It is all well and good to reflect upon the human future and the challenges we will face, but until we experience these possibilities in concrete personal terms, they will not touch us. One way of getting a preview of the coming dilemmas is to imagine how we would deal with them now. This exercise, of course, will only be as useful as we choose to make it. So suspend your disbelief and try to grapple with the following situations as though you are actually confronting them and will experience real consequences from your decisions. Pretend that you really are living in the future. Accept that the procedures are safe and reliable. And remember, there are no right or wrong answers, just ones that are right or wrong for you. So if you need to modify or add to the assumptions in one of the scenarios, to sharpen its issues for you, go right ahead. These questions are tools of self-exploration, and ideally they lead to more questions, not to easy answers.

1. You always thought you were just a normal child, conceived like any other. But you stumble on some papers indicating that your parents picked you from among several hundred embryos they tested, searching for one with your sex and a predisposition for your basic qualities. Would this discovery make you feel better about yourself or diminished? Would it change your feelings about your parents, the way they raised you, and the expectations they have for you?
2. Cloning has come of age, and your favorite pet, which you've raised since its birth, is old and near death. Assuming you have decided to get a replacement, would you clone the animal so that you could start all over again with the same genetics, or would you want a different animal?

3. You are unable to conceive a child and have decided to adopt a baby. You have only two options: you can obtain one from an agency, as was done in the past, or select one that will be conceived from donor egg and sperm and carried to term by a surrogate mother. Which would you choose? The donor list contains accomplished artists and performers, Nobel laureates, movie stars, and many others, each with vital data about themselves. In both cases, the process will take about a year. You can easily afford either option. If you decide to choose the parents, what type would you want? If you choose a regular adoption, would the sex of the baby matter to you?
4. Genetic engineering is illegal in this country, but technically feasible. A routine genetic test on a four-year-old boy injured in a tumble shows that he has an artificial chromosome designed to give him an extra twenty years of good health. The boy's parents are arrested, charged, and found guilty despite their claim that the implantation was their responsibility to their child. Reports show them to be excellent parents. What punishment, if any, do you think is appropriate? Should they lose custody of the child? How would you react to their press statement that they plan on having another child and intend to fly overseas and avail themselves of the procedure again?
5. With hormone treatments, a woman can have a baby after menopause. If you could make the decision, at what age (if any) would women no longer be allowed to get pregnant? Should fathering a child after a certain age be illegal?
6. Imagine that because of an infertility problem, you and your spouse are having a child using advanced *in vitro* fertilization. You have produced a dozen apparently healthy, several-cell embryos and need to pick one to implant. The others will be frozen and stored. Which of these three options would you choose? You can pick one at random; you can use PGD to screen them for genetic diseases and implant one that is healthy; or you can have detailed genetic workups on them and pick one with a predisposition you like. If you choose the full screening, what two traits would be most important to you?
7. If you found out today that your parents could have given you an extra twenty years of healthy life expectancy through a genetic

alteration at the time of your conception, yet chose not to because they felt it was better for you to remain "natural," would you be upset or pleased about their decision? If half of your classmates, work colleagues, or neighbors had this enhancement, would you rather be among them or among the unenhanced?

8. Assume the baby you are planning could be enhanced safely using *in vitro* fertilization and genetic engineering. Would you intervene to raise the child's IQ twenty points? If not, what would you say if your kid grew up and asked you why he or she wasn't as smart as the other kids in class?

Acknowledgments

A BOOK LIKE THIS is never the product of a single mind. In exploring the larger implications of our elucidation of the human genome and our growing control over human reproduction, I have drawn inspiration and ideas from many people. Some of them have spoken to me only through their writings or have stimulated my thinking by some tangential comment in a conversation and are unaware of what they have given me. Others know well that their ideas have helped shape and refine this vision of the human future.

I owe a special debt to John Campbell, who first suggested the idea of a conference on human germline engineering. We had been discussing the implications of emerging developments in genetics and biotechnology for some time, and I would never have taken on the task of organizing the 1998 symposium Engineering the Human Germline without his active collaboration. The event opened up public discussion and provided the impetus for this book. I also owe thanks to Donald Ponturo, a longtime friend, who joined UCLA's Program on Medicine, Technology, and Society to help pull together the symposium and who has been a trusted colleague and adviser. He encouraged me in this and other efforts, offering many excellent ideas and suggestions. Lori Fish, my wife, not only provided invaluable emotional support, she challenged my thinking, urged me to dig deeper, and kept my feet planted on solid ground while I explored tomorrow's possibilities.

Several other people have been a great help with various aspects of the manuscript. Laura van Dam was more than just an editor;

she helped me shape the book and, later, assisted me in refining and achieving my vision of it. Her enthusiasm for the project was unflagging, and I am very grateful for her help. Larry Cooper, my manuscript editor, managed to smooth many an awkward passage. Joe Spieler, my agent, has been a trusted adviser. He helped me tighten my thinking and improve my explanations.

My brother, Jeffry Stock, has always been a valued critic and font of ideas. His detailed comments were especially useful and strengthened the story I have told. Joseph Umphries made some very constructive suggestions. Katherine Fassett tracked down permissions. Whitney Peeling managed the publicity. And Mark Tauss designed a great jacket.

I also thank Gregory Benford, Susan Blackmore, Andrea Bonnicksen, Mario Capecchi, Francis Collins, David Comings, Terrence Deacon, Christian de Duve, Aubrey de Gray, Roger Dworkin, George Ennenga, John Fletcher, Richard Gatti, Robert Goldberg, Wayne Grody, Robert Heller, Christopher Heward, David Heyd, Leroy Hood, Sohail Inayatullah, Eric Juengst, Daniel Kevles, Daniel Koshland, Jr., Christopher Lee, Edward McCabe, Glenn McGee, Hans Moravec, Marty Nemko, Gregory Pence, Leonard Rome, Gerald Schatten, J. William Schopf, David Seifer, Lee Silver, Gary Small, John Summer, Peter Ward, James Watson, and Mark Westhusin. In one way or another, they each contributed helpful ideas, suggestions, or support.

Finally, I want to acknowledge the Greenwall Foundation and the Sloan Foundation for their support of the 1998 conference. I am grateful to the Center for the Study of the Evolution and Origin of Life, the Neuropsychiatric Institute, and the School of Medicine at the University of California at Los Angeles for providing me with so rich an environment in which to work.

Notes

1. The Last Human

PAGE

1 Few have thought through: See Silver, 1997, for a solid attempt to examine the consequences of new reproductive technologies.

Fewer than 1 percent: In 1998, 28,000 babies were born by IVF procedures in the United States out of a total of 4 million births. See www.cdc.gov/nccdphp/drh/art98/section1.htm.

2 No impassioned voices: In the early twentieth century, this was not the case. Many eugenics tracts decried the declining birth rate of the upper classes. "When half, or perhaps two-thirds of all the married people are regulating their families, children are being freely born to the Irish Roman Catholics and the Polish, Russian, and German Jews," wrote Sidney Webb in Britain in 1907 (Webb, pp. 16–17). Even Theodore Roosevelt scolded the upper class for committing "race suicide" (Kevles, 1995, p. 74).

Our "germinal" cells: The egg and sperm that join to form the first cell of the embryo and hence provide the genetic foundation for every cell in the child-to-be are germinal or germ cells, as in the germination of a seed. The term "germline" refers to the hereditary line of these cells.

Procedures on nonhuman primates: See Chan et al., 2001.

4 Human Genome Project: For an overview, see Roberts, 2001.

HAL: The computer in Stanley Kubrick's 1968 film, *2001: A Space Odyssey.*

5 Reject human enhancement: See, for example, Kass, 2001.

Efforts to clone humans: For a good overview of early announcements to clone humans, see Talbot, 2001.

Gene-therapy trials: For a complete list of gene-therapy trials at various stages of review in the United States, see www4.od.nih.gov/oba/rac/protocol.pdf.

6 "Reflect upon the extraordinary advance": Butler, 1872, pp. 199, 203, 207, 221.

7 Patent attorneys at Roslin: In March 1996, the Roslin Institute applied for two patents (GB 2331751 and 2340493), dealing with the nuclear transfer method of creating animal embryos. While the application indicates the intent to use the procedures in "non-human animal" embryos, various claims refer simply to "animal" embryos, so as to cover humans as well. Ian Wilmut and Keith Campbell, as the inventors, necessarily agreed to these claims. This does not mean that they do not have grave reservations about human applications, but it suggests that someone at PPL or Roslin anticipated that various human applications might be important commercially.

8 "We don't say doping": Dahlburg, 2000.
The public will want them: A good example of this is the use and misuse of human growth hormone today.

9 Public approval of aesthetic surgery: Bardo, 1998.
Not only for the wealthy: Siebert, 1996.

10 Look for an edge: The World Anti-Doping Agency sponsored a meeting at the Banbury Center in New York in February 2002 to figure out how to combat genetic enhancement in sports should such technology become feasible (Keating, 2001).
People will worry: Two types of risks are inherent in the regulation of new technologies: the damages that occur when we move too quickly and make mistakes, and those lost might-have-beens that result from excessive caution. Our minds give more weight to concrete threats than abstract possibilities, even if the immediate risks are minor because they would involve few people and the potential later benefits are immense because they would affect millions. We respond emotionally to the thought of a child hurt by an experimental technology, but a patient who dies because of a delayed cure is just as much a victim.
"small . . . societies": Diamond, 1993, p. 336.

11 1998 panel: This was part of the symposium Engineering the Human Germline, which I convened with John Campbell, a colleague at UCLA, at the university's Center for the Study of Evolution and the Origin of Life. For more details, see www.ess.ucla.edu/huge; www.ess.ucla.edu/huge/timesarth.html; or Kolata, 1998.

12 Opposed such interventions: Kass, 2001.

13 "We should never": Winter, 2000.

14 Study in Bombay: Kristof, 1991, and Rao, 1986.
Physician poll: Nippert, 1999.
Female infanticide: In the 1800s, some Sikh villages had sex ratios as low as 31 women to every 100 men (Patel, n.d., and Dickemann, 1979).
Sex selection: A collection of diverse opinions about the various challenges

of sex selection is available in the inaugural issue of the *American Journal of Bioethics,* Winter 2001, pp. 2–39.

Shortage of girls: Self-corrections would likely arise as girls became less common and therefore more favored. For example, in some cultures where girls have better opportunities, parents prefer them. Among the impoverished Mukogodo of Kenya, studies show that baby girls are given better health care because they can marry into the harems of wealthy Maasai and Samburu men, whereas Mukogodo boys cannot escape their poverty (Ridley, 1993, p. 126). Furthermore, if there actually began to be too many boys, ample time would exist to create social incentives to reverse the trend. In China, where boys are more valued than girls, a family is allowed only one child, and girls are taken into their husbands' families, the incentives for gender selection are about as strong as they could be, yet the ratio of newborn boys to girls is only 6 to 5. Nonetheless, the problem in China shows how hard it is to regulate a single facet of reproduction, such as family size, without eliciting compensatory adjustments in individual behavior. For anyone concerned about the long-term social implications of gender imbalances, it is worth noting that following the battlefield slaughter of soldiers in World War I, the ratio of young women to men in England was very high, with only temporary consequences.

16 Larger evolutionary context: Stock, 1993, pp. 2–40.

jellyfish gene: For an art project, Eduardo Kac contracted with a research lab in France to create Alba, a transgenic rabbit containing the gene for a green fluorescent protein found in the Pacific Northwest jellyfish. See Kac's Web site: www.ekac.org/genointer.html.

2. Our Commitment to Our Flesh

19 *Johnny Mnemonic:* Gibson, 1994.

20 End to bacterial infections: Lederberg, 2000.

"Porting our brains": Kurzweil, 1999.

21 Tapping into the brain: Calculating the computing power of the brain is one thing, and even comparing it loosely with the power of our computational devices is plausible, but the conclusion that there will be some easy linkage between the two does not follow. And this is precisely what Hans Moravec imagines. To him, the corpus callosum, the band of 200 million fibers connecting the left and right cerebral hemispheres, would be the ideal portal for establishing such a linkage, and he thinks he is being conservative in saying it will take even fifty years to accomplish. Conversation with Moravec, April 13, 2001.

Cochlear implants: Cochlear Corporation is a leader in cochlear implants and has pioneered a brainstem implant for those who have suffered damage to their cochlear nerves. See www.cochlear.com. For attempts to create visual stimuli for the blind, see Dobelle, 2000.

21 Mechanical heart: Abiomed makes the heart, which weighs about two pounds and is the size of a grapefruit. An external battery powers it, and a rechargeable internal battery, about the size of a pager, operates the device for up to thirty minutes, so the wearer can at least remove the external battery for short periods. Mestel, 2001.

23 Link between brain and computer: I asked several people drawn to the idea of cognitive enhancement what precise brain upgrade they would want. They described abilities they wish they had: solving differential equations in your head; knowing where you are at any time and how to get anywhere; remembering all the music you had ever heard or any conversations you had previously held; having vast stores of knowledge on every subject. But given how easy it will be to approximate these capabilities using miniature external devices, I suspect that no one would subject him or herself to the surgical risks required for implants.

24 Pilotless fighter: Richter, 2000.

25 Fyborgs: See Chislenko, 1995.
We already are fyborgs: See Chislenko, 1995, for a fyborg self-test.

26 Fyborgs and legacy systems: See Chislenko, 1995.

27 Brain implants: Any widespread use of implants for purposes of enhancement rather than repair of compromised functioning would require that surgical trauma be so reduced that surgeons could gain access to the inside of our bodies as easily as the outside. Achieving this would require futuristic technologies such as "cell herding," which would move cells out of a scalpel's path, but such sophisticated biological manipulation would also greatly advance direct biological enhancement, thereby raising the bar even higher for implant technology.
Warwick's implant: His plan is to place his next implant beside his vagus nerve, a bundle of several thousand individual fibers, and pick up its signals, record them, and process them. By later retransmitting the signals to the quiescent nerve, he hopes to recreate movements, sensations, and emotions, but this is a tall order. Warwick, 2000.
"I believe humans": Warwick, 2000.

28 Ankle bracelets: Sophisticated systems of this sort, relying on GPS sensors, have been in use since about 1998. See, for example, the Web site of Protech Monitoring, www.ptm.com. The tracking and interactivity are getting quite sophisticated.
"Just think": Warwick, 2000.
House of the future: Hans Moravec is building the first fully self-navigating robotic cleaning appliance and hopes to have it on the market in about five years. The appliance will avoid furniture and other obstacles, recharge itself,

and empty the dust it collects. But I doubt it will work in the home. Consider the scattered toys, twisted throw rugs, cups left on the coffee table, unmade beds, and messy countertops. Vacuuming the floor covers perhaps a twentieth of the work of housecleaning, and you'd still have to keep your vacuum cleaner for other tasks — from cleaning upholstery to vacuuming the windowsills. Ignoring such complexity goes along with the exuberance about future links with our nervous systems. These problems are far harder than engineers imagine.

29 Wearable electronics: In March 2001, the FDA approved a device called the GlucoWatch Biographer, which measures glucose levels every twenty minutes, extracting test fluid through the skin using tiny electric currents. If glucose levels become dangerous, it sounds an alarm. See *FDA News*, "FDA Approves New Glucose Test," www.fda.gov/bbs/topics/news/2001/new00758.html.

Doubling of computer power: See Kurzweil, 1999 (pp. 103–5), for projections of future computing power. He uses what he calls the "law of accelerating returns" to predict an ongoing doubling in the capacity of neural-net computers. This would lead to $1,000 computers that match the computing speed and capacity of the human brain by around 2020. He continues the projection up to the year 2100.

30 "The fact that materials": Jones, 2000.

Molecular computing: Rotman, 2001.

33 Further gains in longevity: See Olshansky et al., 2000.

Austad believes: He made this prediction at the conference Critical Future Milestones on Aging, in February 1999. See Kolata, 1999, or http://research.mednet.ucla.edu/pmts/aging.htm. Austad later bet the demographer S. Jay Olshansky that by 2150 someone would reach 150 years of age. Each put $150 in an account, which should grow to $500 million by that time.

34 Differences we have created: The historical origins of dog breeds are almost impossible to trace before a few hundred years ago. Ash (1927) provides some information. The American Kennel Club suggests that poodles were water retrievers known in continental Europe in the 1500s, and images of dogs resembling the Great Dane, which was bred in Germany, are supposedly inscribed on Egyptian monuments dating back to 3000 B.C.

3. Setting the Stage

35 Doctors had assured Jesse: See the testimony of his father: Gelsinger, 2000.

36 Whole endeavor in question: Rosenberg and Schechter, 2000.

Consent forms: See Kast, 2000. This is an excerpt from his informed-consent agreement at Johns Hopkins University: "We do not know what the risks are when people are given the altered AAV virus with the CFTR gene. The altered virus could spread to other parts of your body. The consequences of this are not known at this time. There is a very small chance that the altered virus could damage the DNA in the cells of your lungs or nose . . . If this happened, it could put you at risk for developing cancer in the future. You will receive no therapeutic benefit from this. Side effects in humans are not known. If you should die either during or after this study, we will ask your family for permission for an autopsy."

37 "My body produces": Kast, 2000.
Severe combined immunodeficiency: See Cavazzana-Calvo et al., 2000.
A group in Montreal: Kay et al., 2000.

38 "An amazing thing": See Stock and Campbell, 2000, p. 83.
Viral vector: This is a virus altered to carry a payload of DNA into the genome of cells it infects. But the DNA is often inserted at a random location in the genome and the process is not very efficient.

39 Dr. Blaese's clinic: Grady, 1999.

41 Projects to sequence the genome: See Sinsheimer, 1989; Dulbecco, 1986; and DeLisi, 1988. In the summer of 1985, Dulbecco proposed the project as a way of understanding the genetics of cancer, and followed up in 1986 with an influential editorial in *Science* espousing the idea. For a concise review of the origins of the Human Genome Project, see www.fplc.edu/risk/vol5/spring/cookdeeg.htm, and Cook-Deegan, 1994.
Early genome projects: See *The U.S. Human Genome Project. The First Five Years: Fiscal Years 1991–1995,* a publication of the National Human Genome Research Institute. www.nhgri.nih.gov/hgp/hgp_goals/5yrplan.html.
More than thirty thousand: At the time of the June 2000 announcement of a rough draft, the number of human genes had not been determined. But by February 2001, when the long-awaited papers on the genome were published in *Nature* and *Science,* it was clear that the final total would be between 30,000 and 35,000. Most experts had guessed far higher. For instance, the Sanger Center, near Cambridge, England, set up a sweepstakes among gene researchers to estimate the total number of human genes, and by early 2000 had 250 entries. The median was some 54,000 genes, and the range extended from 27,500 to more than 200,000.

42 Breast cancer risks: See Struewing et al., 1997, and "Case Studies in Breast Cancer," www.cancergenetics.org/bc.htm#test-bc.
"This is the big one": See Wade, 2000.

43 Huntington's disease: See Ridley, 1999, p. 56.
Unaware of any genetic: Once we do know our genetic vulnerabilities, we will be more motivated to alter our behavior and take drugs to help protect ourselves. New preventive measures will move medicine quickly toward preventive methods targeted at people at high risk.

44 Differences are cultural: Indeed, some hard-line feminists have even claimed that morning sickness and the pain of childbirth are socially constructed rather than biological. See Patai and Koertge, 1999.
Unraveling genetic differences: Many differences in disease susceptibilities — the prevalence of Tay-Sachs disease among Ashkenazi Jews or of sickle cell anemia among blacks — are known. The probabilities of specific traits within different populations differ as well. First menstruation occurs at a younger age in African populations, and Gregerson et al. (1999) report that perfect pitch is more prevalent in Asian populations.
Genetic diversity form: Olson, 2001.

45 "As a scientist," "It is potentially racist": See Entine, 2000, pp. 29–35, 11, 12.

46 Probe our genetics: In 2000, Orchid Biosciences, in Princeton, New Jersey, for example, stated it would launch a Web site, called shieldgene.com, that would allow individuals to match their DNA with a database of genetic variations. Whitman, 2000.
As little as $200: Personal communications with Jeanne Loring, chief scientific officer, Arcos BioScience, and with Kristi Hulett, Affymetrix Corporation, January 2001.
Working on chips: One method uses tiny dots of probe laid out by ink-jet printers. Another uses tiny capillaries, each filled with a different probe, packed like the hair of a ponytail so that the bundle can be cut into thousands of thin slices, each with hundreds of thousands of capillary cross-sections. Yet another method uses millions of fiber-optic strands, each with little nicks at its end to hold tiny beads that laser pulses within the individual glass fibers can probe.
Gene expression in tumors: Classifying cancers by their patterns of gene expression is now becoming feasible, and this is bound to lead to better ways of choosing the most effective treatment for a particular tumor. Moreover, several cancer centers are exploring ways of creating shared tumor registries.
Obstacles to overcome: Variations in unexpressed sequences and modifications to the long regulatory regions preceding genes, for example, are far more difficult to detect than a change in a gene's coding sequence, which is what is usually discussed in the media. Kan et al., 2000.
Yield enough information: In 1999, there were only about 5,000 genes that we knew anything about, and there were maybe 500,000 papers written on one or another aspect of them. At the rate of a few person-years per paper, that would come out to a couple of million person-years of effort to learn a small amount about a sixth of our genes. In mid-2001, less than three years later, at least 90 percent of the human genome was sequenced, and the best estimates were that there were somewhat more than 30,000 genes to probe and examine. (Conversations with Christopher Lee, a professor at UCLA and a leading expert in bio-informatics, in May 2001.)

47 More than $100 million: In February 2000, Venter's net worth topped $500 million when Celera stock pushed above $200 per share in the excitement over genome sequencing (data from SEC filings: http://biz.yahoo.com/t/03/5948.html).

Single gene in a worm: For some thoughts on the likelihood of common mechanisms for aging in roundworms, fruit flies, and humans, see Strauss, 2001.

Crigler-Najjar: See Kren et al., 1999.

49 Family studies: These will still be valuable, however, because they will lead researchers to groups of individuals who are likely to have revealing combinations of genes. Researchers can do breeding experiments to create these sorts of combinations in laboratory animals, but naturally occurring inbreeding in populations and families is required to create equivalent combinations in humans.

Collecting patient histories: There are even attempts to use the Internet for such collection. DNA Sciences, Inc. (www.DNA.com), reported in 2001 that in its first year it had collected blood samples from ten thousand volunteers with conditions including cardiac arrhythmia, breast cancer, colon cancer, type 2 diabetes, and multiple sclerosis. See Wong, 2001.

50 First transgenic primate: In January 2001, Schatten reported the creation of the first transgenic primate, though it had only a marker gene that could be easily identified in adult tissue. The gene was inserted by retroviral transfer rather than artificial chromosome technology, hence the name ANDi (DNA inverted). See Chan et al., 2001.

For about $7,000: For example, see the Northwestern University fee schedule, www.northwestern.edu/research/shared-facilities/nucmier-tmf.html.

51 Such models are valuable: See Marx (1994) for Alzheimer's, Strober and Ehrhardt (1993) for inflammatory bowel disease, Snouwaert et al. (1992) and Colledge et al. (1995) for cystic fibrosis, and Atkinson and Leiter (1999) for type 1 diabetes.

Birth of Dolly: See Wilmut, 1997.

Successful clonings of mice: See Wakayama et al., 1998, and Prather, 2000. Genetic Savings & Clone intends to clone dogs and cats, charging an initial fee of $200,000. An associated team at Texas A&M, working off a private grant of $2.3 million, is developing the procedure.

Zavos testimony: Vogel, 2001. The full testimony can be found at http://frwebgate.access.gpo.gov/cgi-bin/getdoc.cgi?dbname=107_house_hearings&docid=f:71495.wais, and Talbot, 2001. For an appraisal of the dangers involved, see the testimony of Rudolf Jaenisch, a professor at the Massachusetts Institute of Technology. As of this writing, attempts to clone humans have little chance of success and a great chance of creating abnormal fetuses if embryos are implanted. But much effort is going into elucidating the reprogramming of genes during cloning procedures in cattle and mice, and I would expect someone to clone a human during this decade.

52 Pharmaceutical-laden sheep: See Lewis, 1998, pp. 1, 7.
Extracting clotting factors: See Paleyanda et al. (1997) for a discussion of treating hemophiliacs with human factor VIII from pigs, and see Schnieke et al. (1997) for a discussion of human factor IX from sheep.

53 "Monstrous . . . What are the psychological": See Pence, 1998, p. 26; Rifkin and Howard, 1977, p. 15.
28,000 babies: See www.cdc.gov/nccdphp/drh/art98/98nation.htm. The U.S. Center for Disease Control keeps track of this information. Approximately 40 percent of cycles led to pregnancies, 80 percent of those led to live births, and of the live births, 61 percent were singletons, 28 percent were twins, and 11 percent were triplets.

54 Headline stories: See Stourtin, 2000, and "How Old Is too Old?" BBC Online, July 16, 1999. Severino Antinori, the same doctor now planning with Panos Zavos to clone a human, runs the Italian clinic that treated della Corte.
"Tiegs underwent": See Dam and Wihlborg, 2000.
By forty, her chances: For success rates in older women using their own eggs, see www.cdc.gov/nccdphp/drh/art98/images/fig10.gif. The likelihood of pregnancy using donor eggs, however, remains about 50 percent per cycle, roughly the same as for a woman in her twenties, the age of most donors. Each of the half-dozen IVF doctors I asked about the possible truth of the Cheryl Tiegs story agreed that it was not believable. For a larger discussion of IVF in older and postmenopausal women, see Fisher and Sommerville, 1998. Note that natural births to older women occasionally do occur. In France, about 1 in 70,000 births in 1990 were to women fifty-two or older, but the likelihood of fraternal twins would be much lower.

56 Combine to convert IVF: Edirisinghe (1997) discusses *in vitro* egg maturation; Gosden et al. (1993) examine follicle maturation; Candy et al. (2000) discuss ovarian grafts; Kuleshova et al. (1999) discuss egg vitrification; Wood et al. (1997) discuss cryopreservation; and Isachenko et al. (2000) review the field. An eleven-week-old fetus of a baby girl already has the full complement of 1 to 2 million oocytes that will constitute the growing egg follicles in the sexually mature woman (a follicle refers to an egg and its surrounding cells). Today, freezing ovarian tissue can preserve these small follicles, and when the tissue is later thawed and grafted to a recipient ovary or vascularized under the skin, the follicles will mature. Alternatively, once follicles in the ovaries reach a diameter of 3 to 10 millimeters, they can be extracted and matured *in vitro* to produce a viable pregnancy. Bridging the gap between these two procedures is the challenge.

56 First performed PGD: See Handyside et al., 1990, and Handyside et al., 1992.

Future use of PGD: Testing for a single genetic mutation on a single cell is possible today. With a child at risk for cystic fibrosis, for example, a lab first tests the parents to determine which particular mutation might pass to the child. Researchers have cataloged nearly a thousand distinct mutations that can cause this well-studied disease. According to Grody (1999), in the Caucasian population, the most common one occurs in 70 percent of sufferers, but there are some twenty more that occur in about 1 in 1,000 sufferers. To test for a few markers at once using a single cell is feasible today using multiplexed PCR, a technique to amplify each sequence that is to be tested, but what is known as whole genome amplification is improving rapidly and could make it possible to test directly for hundreds or thousands of gene variants on a single cell. Today's simple PGD tests are easy to complete within twenty-four hours, which means that embryos do not have to be frozen until results come back, and the same will likely be true for advanced PGD.

Parents will have a choice: See Silver, 1997, pp. 233–65.

57 Genetic contributions: See Mestel, 2000. Studies of twins suggest that genetics can account for about a third of cancers, but this influence varies by cancer type. Of common cancers, prostate cancer seems the most highly linked to genetics, with more than 40 percent of incidences correlated.

Environmental influences: See Sapolsky, 2000. When a mother rat licks and grooms her baby, she initiates a cascade of responses in the pup that eventually turn on its growth genes.

58 International poll: I cite the figures of Macer et al. (1995, p. 798) on physical enhancement, but the figures for added intelligence are similar.

Millions of couples: See Feldman, 1998.

Less expensive: Generally, insurance does not cover IVF, but the state of Massachusetts mandates insurance coverage for infertility, and this seems to cause a significant increase in the use of the technology. See http://apps.nccd.cdc.gov/art98/clinics98.asp.

4. Superbiology

62 At age seventy, about half: See Struewing et al., 1997; www.cancergenetics.org/bc.htm#test-bc.

63 Task facing genetic engineers: The success of present efforts to unravel polygenic diseases and conditions is not a good guide to what will happen in the future. Today, geneticists work on conditions that are of particular interest and are fundable, but when broad gene surveys are possible, our genomes themselves will tell us what conditions look most ripe for exploration.

Genetics of perfect pitch: See Baharloo et al., 1998 and 2000.

65 Close to foolproof: One might imagine that it would be good enough for germline interventions to work as well as natural conception. For example, some 8 percent of us have some sort of genetic abnormality that shows up by age twenty-five. See Baird et al., 1988. I suspect, however, that litigation associated with failures in an elective technology of this sort would render mistakes so expensive, at least in the United States, that the procedures will have to be much safer than natural ones to be widely used. It's hard to sue Mother Nature.

Human artificial chromosomes: See Harrington et al., 1997; Willard, 2000; and Choo, 2001.

Synthetic chromosomes: See Willard, 1998.

Successive generations of mice: See Coghlan, 1999, and Deborah, 2000.

Would not license: See Dove, 1999.

67 Finely tuned genetic controls: See Misteli, 2001.

Chromosome 21: See Hattori, 2000.

68 "Silence" the original gene: One way of inactivating a gene is to add another gene that makes a reverse or "antisense" sequence of the original gene's messenger RNA (the molecule that carries a gene's message from the chromosome to the place where the cell fabricates its proteins). The antisense sequence binds the messenger and keeps it from doing its job of transcribing the gene into protein. Even with a little leakage, this would effectively silence the original.

Banding patterns: See Yunis et al., 1980.

69 Tetracycline as the signal: See Kitamura, 1998; Gao et al., 1999; and Harvey et al., 1998.

70 Creation of knock-out mice: See Capecchi, 1989.

Keeping a chromosome from passing: See Capecchi, 2000.

"Unnecessary to save": See www.gene-watch.org/programs/Position_Germline.html.

71 Reaffirm their parents' vote: One might say no option exists if it means coming down with some horrible genetic disease that the parents had therapeutically avoided. But this was the same choice that faced the grandparents.

Enzyme called CRE: See Sternberg and Hamilton, 1981.

Remove a test gene: See Bunting et al., 1999.

Targeted deletion in fruit flies: See Golic and Lindquist, 1989, and Golic and Golic, 1996.

72 Engineered T cells resistant: See Duan et al., 1997; Biasolo et al., 1996; and Woffendin et al., 1996.

73 CD4 is a marker: See Browning et al., 1997.

74 Ecdysone, an insect hormone: Christopherson et al., 1992.

Tetracycline: The "tet" system is one of several now in use to activate genes conditionally. It has been well studied and widely used, and can turn on a gene in as little as one hour; the process can be reversed. See Spinney, 2000.

75 Markers would make it easy: See Vassaux, 1999, and Narayanan et al., 1999. Allow better control: Young et al., 1999.

77 Gets a new shuffle: Every adult has two copies of each chromosome — one maternal, one paternal — and thus two copies of each gene. Each sperm or egg, however, has only one copy of each gene. The gene mixing that occurs, making every sperm or egg unique, is called crossing over, and on average it happens two or three times for each chromosome. Thus there is a great deal of mixing between generations.

5. Catching the Wave

79 Would not reach eighty-five: See Olshansky et al., 2001. The life expectancy gains are small because reducing age-specific death rates by equivalent percentages brings an ever smaller increase in lifespan as the survival rates rise.

80 Compress our final decline: For a fuller discussion of this, see Gems, 2002.
Research into the biology of aging: See de Grey et al., 2001.
True aspirations for medicine: If we are in general physical decline and every aspect of our physiology is gradually running down as we age, then by avoiding death from specific age-related conditions such as atherosclerosis, we simply ensure that the decline in all our bodily systems is relatively balanced and that by the time we die, our bodies will be failing throughout. Many people find this picture of general frailty and progressively diminished functioning distinctly unappealing, yet it is inherent in the existing goals of medicine.

81 Roundworm lifespans: See Kenyon, 1996.
Increased mouse lifespans: See Weindruch and Walford, 1988.
Some 25 percent: See Herskind et al., 1996, and Finch and Kirkwood, 1999. Our fetal environment, mere chance during our earliest cell divisions, and other largely uncontrollable happenings contribute greatly to the rate at which we age.
Sardinian centenarians: See Koenig, 2001. Genetic studies are under way to look at every adult in several small villages in Sardinia that show elevated longevity. Perhaps ten thousand villagers are in the studies.

82 Variation is still great: See Weindruch et al., 1986.
Similar effects with primates: See Pugh et al., 1999.
Internet group: See www.infinitefaculty.org/sci/cr/crs.
Pathways involved in postponed aging: See Lee et al., 2000, and Lee et al., 1999.

83 October 2000 roundtable: Strategies for Engineered Negligible Senescence (SENS): Retarding Rather than Reversing Aging. See http://research.mednet.ucla.edu/pmts.

84 Role of micronutrients: See Ames, 1998.
Reversing aging: See de Grey et al., 2000, and http://research.mednet.ucla.edu/pmts/sens1.htm.

85 Implications of germline therapy: We must keep in mind that the alternatives of a full realization of an anti-aging intervention in adults or only in embryos represent the two extremes. Many dozens of genes may be associated with various aspects of aging, and many nongenetic influences as well. Specific dietary changes, pharmaceutical regimes, and genetic interventions may each achieve aspects of what only caloric restriction can accomplish today. The optimal mix of interventions will depend on an individual's age and genetic constitution as well as the current state of medical science.

87 Aging as *the* disease: Huntington's disease is a fatal condition that evolution cannot purge because it acts late in life, after reproduction has occurred. Late-acting, mildly deleterious conditions accumulate in the human genome for the same reason. One way of viewing aging is as a diffuse, late-onset genetic disorder, the result of a host of genetically based conditions that manifest themselves as we get older. Aging is a disease in the sense that it is the critical underlying factor in all the debilitating diseases of advancing age — what makes us "old" and brings on these diseases — so regardless of which bodily system is the first to deteriorate and break down, causing death, aging is often the root cause. Temple et al. (2001) suggest that in the genomic era, our thinking about disease will change dramatically, and that we now need to view disease not as a cluster of symptoms but as "*a state* that places individuals at *increased risk* of *adverse consequences*." The authors do not mention aging, but the process clearly embodies all three elements (in my italics) of their definition of disease.
Aging of the workforce: Without the development of technologies to extend our vital years substantially, various countries with rapidly aging workforces will soon face the challenge of maintaining their standards of living. This will be especially difficult for Japan and Germany, which seem culturally resistant to the influx of young foreigners who may be required to sustain them. Both nations have recently witnessed political backlashes against increased numbers of immigrants, from Korea and eastern Europe respectively. Japan's fertility rate has fallen to 1.33 children per woman. At this rate, its population will fall from the current 127 million to some 88 million in 2050. Moreover, 40 percent will be over sixty-five and 18 percent over eighty. See Effron, 2001.

88 Psychological tremor: Little excitement greeted the doubling of fruit fly and roundworm lifespans, but these animals live a matter of weeks, not years, and the results have less obvious relevance to humans. Mice are mammals and our close genetic cousins.

89 $30 billion a year: Given how lucrative a specialty plastic surgery has be-

come and the many thousands of people willing to spend $5,000 or more a year on human growth hormone supplements, I suspect that people would be willing to pay a lot for something that really did slow aging and make them feel younger.

90 Drug approval process: See Zivin, 2000. Today's clinical studies are the gold standard of testing. Nothing beats a double-blind, fully controlled, long-term study in which neither patients nor doctors know what is being administered. But such studies cost too much and take too long for conditions like aging.

93 ICSI: See Palermo, 1992.

94 Lemurs as research animals: The most common monkey in research is the rhesus, which lives an average of twenty-five years and sometimes survives to the age of forty. This is too long for testing anti-aging interventions. Incidentally, the type of parallel test that may one day be used for germline interventions is under way today for IVF. At the Wisconsin Regional Primate Center, Petri, the oldest rhesus conceived by IVF, is still going strong: he turned seventeen in August 2001.

96 Increase in standard of living: See Nordhaus (1999) for calculations on standard of living. See also Murphy and Topel (1999) for an additional discussion of this issue.

6. Targets of Design

98 Headlines about gene discovery: See Ritter, 2000: "Scientists Find Fat Gene"; Siegel, 1996: "Israeli, U.S. Scientists Find Risk Gene"; Tamkins, 1999: "Single Gene May Send You to Bed Early"; Verrengia, 1999: "French Locate Seizure Gene"; and Kmietowicz, 2001: "Genes Implicated in Vision Problems."

Aspects of their personalities: Such information comes from studies of identical twins raised apart from birth. These studies suggest, for example, that the amount of leisure time devoted to religious activities — that is, time spent in a house of worship or reading religious texts — is about 50 percent heritable. See Wright, 1999.

Sociobiology: See Wilson, 1975 and 1978, and Wright, 1994.

Gould attacked Wilson: See Wolfe, 1999. In the *New York Review of Books,* Gould warned of the danger of such theories and asserted that they tend to "provide a genetic justification of the status quo and of existing privileges for certain groups according to class, race, or sex." He also stressed that in the past such theories had provided "an important basis for the enactment of sterilization laws . . . and also for the eugenics policies which led to the establishment of gas chambers in Nazi Germany."

99 Minnesota twin study: See Bouchard et al., 1981 and 1990, and Bouchard, 1994.

100 Pseudo-twins: See Segal, 1999, pp. 152–68.
Hundreds of studies: Some critics, like Leon Kamin, have pointed to shortcomings in twin studies, such as the self-selection of gay subjects, who are sought through advertisements in gay magazines. While such criticisms may be valid for particular studies and traits (the data are not easy to collect), they do not change the overall implications of this large body of work. See Wright, 1999, pp. 67–84. See www .nomotc.org/research .htm for a posting of some twenty-five twin studies that are seeking volunteers.
Environmental influences shared: Shared genes, for example, are thought to be about five times as important as shared environment in influencing personality traits in identical twins, and about 50 percent more important than unshared environment. Steen, 1996, p. 173.
Tend to be closer in IQ: See McCue et al., 1993, and McCartney et al., 1990.

101 Antisocial behavior: See Bohmann et al., 1982. The genetics of antisocial behavior is even more age-dependent than that of IQ. Genetics seems to account for little of the variation in juvenile troublemaking and crime, but perhaps 40 percent of that in adults.
Challenged as flawed: See Joynson, 1991.
Between 45 and 75 percent heritable: There are several ways of estimating how much our genetics determines our IQ. One is to look at identical twins reared apart. Another is to look at the differences between identical twins who have been reared together. See Hamer and Copeland, 1999, pp. 218–19.
Adopted children living together: See Scarr and Weinberg, 1978, and Loehlin et al., 1989.
Influence of living in one home: See Scarr, 1992.
People who score higher: See Jensen and Munro, 1970, and Neisser et al., 1996.
Most useful predictors: Tambs et al., 1989. There are many determinants of professional success that IQ does not measure — creativity, persistence, energy, sociability, health, and plain old luck, to name a few — but IQ tests are nonetheless very useful in forecasting success and earnings. See, for example, Jensen, 1998.
Jim twins: See Wright, 1997, pp. 43–54. Thomas Bouchard inveigled some grant money from the University of Minnesota and persuaded the two brothers to undergo a battery of tests.

103 More frequently in identical: This observation was made by twin researcher Nancy Segal in a personal communication.
One twin gets Alzheimer's: Wetherell, 1999, and Viitanen, 1997.
May inhibit Alzheimer's: The untangling of effects such as this will drive

our exploration of our own genetics and usher in a broad shift toward preventive medicine. As we identify our risks, we will not sit idly by; we will want to do something to reduce them.

103 Genetics can explain: For height and weight, see Steen, 1996, pp. 161–83; for autism, see Rutter et al., 1993; for schizophrenia, see Wright, 1999; for bipolar disorder, see Bertelsen et al., 1977. For extreme manifestations of personality traits, heritability seems to be much higher than for milder manifestations. For example, extreme shyness is 70 to 80 percent heritable, narcissism about 65 percent, and self-contempt about 60 percent. See Steen, 1996, and Hamer and Copeland, 1999.

104 From 40 to 60 percent: See Bouchard, 1994. Steen, 1996, has a good discussion of personality testing, heritability, and twin studies. Additional material is in Hamer and Copeland, 1999.

Religion is important: See Steen, 1996, p. 170, and Wright, 1999.

105 Such politically charged areas: In 1992, for example, the NIH had already approved funds for a conference titled Genetic Factors in Crime: Findings, Uses, and Implications, which was to be held at the University of Maryland and sought to examine the use of biochemical markers and drug treatments for violent and antisocial behavior. But in response to sharp criticism, then NIH director Bernadine Healy withheld the funds, and the meeting was postponed indefinitely. See Horgan, 1993.

Ensure individual privacy: It is possible, for example, to strip medical records of personal information. "Deidentified" records are adequate for the survey research needed for gene association studies and will fall within the constraints embodied in the privacy provisions of the Health Insurance Portability and Accountability Act regulations, finalized in 2001.

106 A "gay" gene: See Hamer et al., 1993.

Findings were soon challenged: See Rice et al., 1999.

Subsequent study was criticized: See Hamer and Copeland, 1999, pp. 197, 198.

ApoE and Alzheimer's: Steen, 1996, p. 156.

Measuring novelty-seeking: See Epstein et al., 1995, and Benjamin et al., 1996.

107 How many genes: For a discussion of the genetics of both novelty-seeking and anxiety, see Hamer and Copeland, 1999. A comment on the number of genes is on p. 46. For a few less complex traits, the number of genes involved has been estimated. For eye color, three to five genes. For skin color, the same. Some traits, such as the ability to curl the tongue, are determined by only a single gene.

Propensity for harm avoidance: See Lesch, 1996, and Hamer and Copeland, 1999.

Shift propensities dramatically: When someone is extremely tall, devout,

aggressive, slow, or shy, his or her offspring will automatically tend to be more moderate in that trait. This drift toward the average occurs because the many factors, both environmental and genetic, that have aligned to produce an exceptional talent or deficit tend not to remain equally aligned in the next generation. Selecting for embryos that are less extreme in their potentials will thus be easier than reinforcing stronger traits. The practical implication is that avoiding deficits in traits will be much easier than amplifying high performance, which may require both embryo selection and genetic manipulation.
Made the mouse more vole-like: See Young et al., 1999.

108 In the roundworm: The genes age-1 and daf-2, for example, are both involved in an insulin-like signaling pathway. Inactivating either one doubles lifespan; inactivating both further increases it. See Vanfleteren and De Vreese, 1995.
Enhances learning and memory: See Tsien, 2000.
Incidents of multiple births: See Galloway et al., 2000.
Ten times as well: See Montgomery et al., 1998.

109 To do more than guess: Various traits are strongly influenced by modifying only a few genes. Indeed, in both animal and plant breeding, traits can be manipulated in ways that wouldn't be possible if networks of genes were too tangled. To really know whether genetic enhancement is feasible, we will have to wait for the results of the next decade's transgenic manipulations of animals and the harvest of genetic association studies on highly heritable human traits. An examination of the dog genome would be very informative as well, because so much variation exists among dog breeds and yet the genetic differences are not dramatic.
"The biggest ethical problem": Stock and Campbell, 2000, p. 79.
"There will come a time": Lander, 2000.

110 Germinal choice: In the late 1960s, a number of geneticists, including Ernst Mayr, J.B.S. Haldane, and Francis Crick, the codiscoverer of the structure of DNA, voiced support for an idea of "germinal choice," based on couples deciding to reproduce by artificial insemination with sperm from particularly gifted individuals (Kevles, 1995, pp. 261–64).

111 Role of chance: For example, one of the two X chromosomes a female inherits becomes inactive at an early stage in the development of the embryo, apparently at random. So women have some patches of cells that express the X chromosome from their fathers, and other patches that express the X chromosome from their mothers.

112 "A coding variation": See Whybrow, 1997, pp. 87, 88.
Safe Choices: Trying to avoid these dilemmas by blocking the possibilities of genetic manipulation might seem appealing. But we already confront these choices when we give a child Ritalin, abort a fetus with Down syndrome, or

inoculate against disease. Without suffering from polio, Franklin D. Roosevelt might not have achieved what he did, but few parents shun the Salk vaccine.

113 Lee Silver considers: See Silver, 1997, p. 279.
"I would oppose germline": See Hubbard, 2000.

114 Heritability of farsightedness: See Hammond et al., 2001. Note, however, that the broad changes in environment and lifestyle brought by technology have greatly increased the incidence of farsightedness.
Choose black skin: Rickets occurs when not enough sunshine penetrates the outer layers of the skin to make vitamin D, which is critical to bone formation. The reduced skin pigment of human populations that have long inhabited temperate climates protects against rickets — in the absence of adequate nutrition — but increases the skin's susceptibility to sun damage.

115 Three to five genes: Brum et al., 1994.

117 Evolutionary past speaks to us: Wilson, 1978; de Waal, 1998; Diamond, 1993; and Ridley, 1999, are particularly interesting explorations of this subject.
More aggressive and violent: Edwards (1987) explores the large feminist literature concerned with male violence against women, but far more male violence is directed against men. See Daly et al., 1988.

118 Soften features: See Thornhill and Gangestad, 1993.

119 "He has wavy": See Schmidt and Moore, 1998.
Social interactions: See Ridley, 1993.
Genetic markers: Testing new genetic constructs that enhance intelligence is going to be very difficult, because no good animal models exist for testing subtle aspects of human cerebral functioning. This problem, however, will not plague interventions that copy genetic constellations identified in highly intelligent individuals. Nature has already done these tests.

120 Dozens of antidepressants: For a list, see Whybrow, 1997, pp. 261–67.

121 Memes: See Blackmore, 1999.

122 Biological technology: The first lesson from technology is that we have not been very successful at discerning either the patterns or the directions it will take. For the most part, experts did not anticipate the Internet or the personal computer. The second lesson is that we have not been able to plan technological advances, predict the larger implications of specific technologies such as television and radio, or block key technologies. The U.S. government's attempt to squelch encryption technology failed miserably. Nonetheless, adopting standards has been helpful, because it channels research and development activity.

7. Ethics and Ideology

125 "We live in America": See Fenwick, 1997, pp. 168–72.
Nash family: See Frankenfield, 2000.
Senator Thurmond: "The National Institutes of Health have led the way in medical research and must be actively involved in stem-cell research," said Thurmond. "I look forward to working with my colleagues to ensure adequate funding and appropriate government support is available to our national medical research community" (American Society for Pharmacology and Experimental Therapeutics, 1999). By July 2001, when President George W. Bush was deciding whether to relax the ban on using federal funds for research on embryonic stem cells, the promising field had moved far enough that many other abortion opponents, Senators Orrin Hatch and William Frist among others, were openly advocating funding.

126 "Borrowing the actual makings": See Fenwick, 1997, pp. 209–14.
"Proper marital act": A senior Vatican official wrote to a fertility specialist in Argentina in 1984 that obtaining sperm using a perforated condom was acceptable from a "moral point of view" because it constituted a "real and proper marital act" as long as there was an actual possibility of insemination in the judgment of the researchers. (Courtesy of Dr. Juan Calamera and Dr. Santiago Brugo Olmedo of Buenos Aires.) However, a recent paper by the National Council of Catholic Bishops rejects IVF entirely. www.nccbuscc.org/prolife/tdocs/part2.htm.
Germany moderated: See Genillard, 1993. In 1993, Gunter Rexrodt, the economic minister, called the changes proposed to the German parliament to remove the time-consuming and costly requirement of obtaining written permission for every genetic experiment a necessary step in ensuring the competitiveness of German pharmaceutical companies.

127 "Both nations and parents": See Gardner, 1995.
Diversity of attitudes: See Macer et al., 1995. In an international survey, from 20 to 80 percent of people in every country polled claimed to be willing to employ genetic engineering to improve the physical or mental capacities their children would inherit.

128 "We are compelled": Kass, 2001.
Right to unaltered genetic heritage: See Fletcher, 1994, and Bonnicksen, 1994. In 1982, the Parliamentary Assembly of the Council of Europe, for example, asserted a "right to a genetic inheritance which has not been artificially interfered with," and "respect for the genetic heritage of mankind."

129 "Cloning must be banned": This quote was taken from a debate between Glenn McGee and me on cloning and germline manipulation. For its full text, see *Wired* magazine's "Hotwired" archive: http://hotwired.lycos.com/synapse/braintennis/97/37/index0a.html. Note that this comment of

McGee's came before the shortcomings of current procedures and the medical dangers attending them became clear.

129 Unwillingness to acknowledge: Those who seek to confound our choices make them even more difficult. Bailey (2001), for example, reports that Martin Teitel, the director of the Council for Responsible Genetics, admitted using such a strategy with the so-called precautionary principle — the seemingly reasonable idea that new technology should be banned as dangerous until it is proven safe. Bailey asked Teitel how any scientist could prove that a biotech crop was completely safe without the field trials that the precautionary principle would simultaneously require and ban. Teitel replied, "Politically it's difficult for me to go around saying that I want to shut this science down, so it's safer for me to say something like 'it needs to be done safely before releasing it.'" This exchange could as easily have referred to genetically modified people as biotech crops.

130 Testing reveals cystic fibrosis: See Grody, 1999.

131 "God gave us molecules": See *Religious Perspectives on Human Cloning*, http://pewforum.org/events/0503, May 2001.

132 "The idea of natural law": Stock and Campbell, 2000.

133 "I do not find": Freundel, 2000.

134 Beginning of human life: This same debate is being played out with embryonic stem cells and medical research. Christian critics have argued that experimentation on such cells is immoral. See Kornblut (2001) and the statements of Abdulaziz Sachedina (professor of religious studies, University of Virginia) and Rabbi Moses Tendler at the May 3, 2001, conference Religious Perspectives on Human Cloning (http://pewforum.org/events/0503).

"The risks of the technique": See Wivel and Walters, 1993.

138 Early eugenics: See Kevles, 1995.

The use in warfare of smallpox: See, for example, Finkel, 2001.

139 Medical errors: See Napoli, 2000. The Institute of Medicine estimated that up to 98,000 preventable deaths occur from medical errors in the United States annually. Even if this estimate is too high, as many have asserted, it dwarfs the potential consequences of cautious early use of advanced reproductive technologies.

140 DES: See Hatch et al., 1998.

141 Steptoe and Edwards: The birth of Louise Brown, on July 25, 1978, was not the first attempt at *in vitro* fertilization. Steptoe, of Oldham General Hospital in London, and Edwards, a physiologist at Cambridge University, had created some eighty previous pregnancies, but none had progressed to term.

Only 8,000 IVF procedures: See De Witt, 1993.

142 The diet drug fen-phen: See Mark et al., 1997, and Brody, 1997.

143 Protection against malaria: Without a massive undertaking to remove each

copy of the hemoglobin S allele, which causes sickle cell anemia when it is homozygous (two copies), the frequency of the allele would not decrease. Moreover, given that individuals with a single copy of the S allele combined with the normal allele are healthy and resistant to malaria, there would be no reason for parents to embark on such an effort. Indeed, if malaria were a serious health problem, it would make sense to intentionally put a single copy of this gene in a child's genome. With genetic screening to protect against homozygosity, the entire population (including those not of African ancestry) could have such a gene as a malaria preventive if it seemed valuable, and they would still not risk having children afflicted with sickle cell. Such a strategy would be impossible without genetic screening, since a quarter of all babies born to heterozygous parents — those with one copy — would have sickle cell anemia. Today, some 9 percent of the African-American population have one copy of the gene.

144 "The lottery could be 'rigged'": Parens, 2000.

145 Before the *Roe v. Wade* decision: See Council on Scientific Affairs, 1992, and Dworkin, 1996.

Cheap enough for everyone: Ironically, if genetic enhancement technology were available to all, it might reduce population differences in important attributes. Genetic enhancement is more likely to improve the health of someone whose family has a high risk of heart disease than someone whose relatives are centenarians, for instance.

Government regulations now exist: The Health Insurance Portability and Accountability Act mandates procedures for all healthcare providers and their partners and agents to protect the privacy of electronically encoded health information. Hefty fines of up to $25,000 per violation ensure compliance.

146 Genetic discrimination: A frequently cited case is that of Terri Seargent, who was diagnosed with the genetic condition Alpha-1 and fired. In her statement to the Senate Health, Education, Labor, and Pension Committee in July 2000, she stated that she was the victim of genetic discrimination: "In April 1999, I went to the doctor for what I thought was an allergy problem. In disclosing my family history I told the doctor about my brother dying from complications due to Alpha-1 at age 37. A test was immediately ordered to see if I too might have the disorder. When the test came back positive, my doctor told me about the replacement therapy that would keep symptoms at bay and protect my lungs against infections, and allow me to live. I was employed for 3½ years where my annual reviews were excellent and I had salary increases of 60 percent. The company was partially self-insured. My preventative therapy treatments began in October 1999 at a cost of $3,800 per month totaling $45,600 a year. After another great review and a salary increase in November by the president and vice president, I was

unexpectedly told my services were no longer needed 5 days before Christmas." The interesting thing about this case of so-called genetic discrimination is that the company did not fire her because of a genetic risk factor, but because her medical expenses were so high. Although her condition was genetic, this was not an issue of genetics but of health insurance and employment practices. http://www.geneticalliance.org/PDF/2000 alerts/dec00.pdf.

147 Genetic privacy: See Marshall, 2001.

149 Consensus will come very slowly: Can we really expect anything else when there is still no consensus on such things as whether it is best to let your infant sleep with you at night? In September 1999, the Consumer Product Safety Commission issued a controversial warning that allowing babies to sleep in adult beds puts them at risk of suffocation or strangulation. See Ferber, 1986, and Sears and White, 1999.

151 Slippery slope: See McGleenan, 1995.

"Positive eugenics": Will, 2001.

Slippery sidewalk: An indication that this is not a slippery slope is the eugenics movement of the early twentieth century. Forced sterilization and other practices drew wide support in many countries, but we turned away from such practices and from eugenics in general. See Kevles, 1995.

152 Deal with its growing power: PGD, though no more controversial than routine prenatal diagnosis, is poised to bring parents a host of discomforting choices about the predispositions and vulnerabilities of their children. And it is merely a prelude to germline engineering and conscious human genetic design.

8. The Battle for the Future

153 Embryo Protection Law: This law makes it a crime to manipulate the genetic information of a human germ cell, use genetically manipulated germ cells for fertilization, create or transfer genetic information identical to that of another embryo, fetus, or any living or deceased person, or create "animal-human" hybrids. In addition, each cell in an embryo is to be treated the same as an entire embryo, which rules out PGD. See Bonnicksen, 1994, and Embryo Protection Law, December 13, 1990, reprinted in *International Digest of Health Legislation* (1991), pp. 60–62.

154 AAAS committee: See Frankel and Chapman, 2000.

Cytoplasmic transfer: Erik Parens and Eric Juengst, two members of the AAAS committee, wrote an opinion piece arguing that cytoplasmic transfer, though apparently safe, should be halted because it represented backdoor human germline engineering and was occurring without sufficient debate. See Parens and Juengst, 2001, and Barritt et al., 2001.

Liability risks: While couples might conceivably sign away their own right to sue if problems arose, they could not prevent a future suit by a child damaged during a flawed procedure. Companies will more likely hesitate to offer a safe procedure than jump to use an unsafe one. Indeed, in 2001 a suit brought by a couple in France for wrongful birth of a child with Down syndrome won substantial damages, reinforcing a $100,000 award in France a year earlier to a child with another severe disability (reported by the BBC, November 28, 2001).

155 "We are repelled": See Kass, 2001. See also Bailey, 1998.
Eugenic rhetoric: See Kevles, 1995.

156 Neo-Luddites worried: See Bailey, 2001.
International poll: See Macer et al., 1995. To prevent fatal diseases, the numbers were even higher: 80 to 96 percent of people said they would use these technologies for this purpose.
They announced that a wealthy: See Pickrel, 2001, and Talbot, 2001.
Aliens had visited him: Ironically, Raël claims that they came to Earth because ethics committees prohibited them from doing the clonings on their home planet. The Raelian beliefs seem to be a spinoff of the doctrines of fundamentalist Christians who believe that humans were created by God a few thousand years ago. The Raelians assert that the ancients mistook the identity of the powerful extraterrestrials who created us, thinking them gods. See www.rael.org, and www.rael.org/int/english/embassy/embassy.html.

157 Raelian cloning project: Nuclear transfer procedures are neither efficient nor safe at this point. While the rate of successful transfers has improved since the time of Dolly, and we know more about human reproduction than about that of any other mammal, problems with unpredictable gene expression are still serious (see Jaenisch and Wilmut, 2001). The science is not yet good enough for any degree of safety, but when genetic reprogramming issues are resolved and the rates of abnormalities begin to fall in animals, a human cloning may not be far off. Given current problems, however, sophisticated preimplantation genetic diagnosis may well precede safe, reliable cloning procedures for humans.
Congressional hearings: See http://energycommerce.house.gov/107/action/107–5.pdf. In July 2001, the House of Representatives passed an anti-cloning bill that would make human cloning a crime punishable by ten years in prison.
Led to birth defects: See Jaenisch and Wilmut, 2001, and Humphreys et al., 2001. Note, however, that a recent review of all published mammalian clonings indicated that the majority of live births have no detectable health problems.

158 Transhumanists: See Regis, 1990.
Board of directors: Marvin Minsky is sometimes referred to as the father of AI. Roy Wolford, a professor of pathology at UCLA, has been a leader in re-

search on the biology of aging since the 1960s, when he demonstrated that caloric restriction could extend the lifespan of mice. Ray Kurzweil, an entrepreneur, inventor, and author, is the winner of the 1999 National Medal of Technology. He built the first music synthesizer and the first text-to-speech reader for the blind.

158 Well-known figures: Judy Campisi is head of the Department of Cell and Molecular Biology at Berkeley National Laboratory and group leader of aging and cancer research. Calvin Harley is chief scientist at Geron Corporation. Cynthia Kenyon is a professor in the Department of Biochemistry and Biophysics at the University of California at San Francisco. See Alexander, 2000.

"Mother Nature": See More, 1999.

"Operating manual": In his last public lecture, the late Douglas Adams, the satirist and author of *The Hitchhiker's Guide to the Galaxy,* asserted that the clearest sign that a technology does not work is that it comes with an operating manual. Technologies such as electricity don't need them. (Personal recollection from the Applied Brilliance Architectural Conference, Sedona, Arizona, May 2000.)

159 One-child-per-family policy: China launched its *wan-xi-shao* (later-longer-fewer) program in 1971. Without it, the country's population at the turn of the millennium might have been 2 billion instead of 1.2 billion.

China's 1995 law: See O'Brien, 1996.

160 Over one hundred varieties: See Smith, 2000.

IVF clinic in Xi'an: The clinic was described to me in late 2000 by David Seifer, the director of the University Center for Reproductive Endocrinology at the Robert Wood Johnson Medical School in New Jersey.

Some forty IVF clinics: Numbers of this sort are difficult to validate; this figure comes from estimates given by several knowledgeable sources. The majority of the laboratories are probably about ten years behind in technology, with a handful, like the one in Xi'an, that are more advanced. As to the commitment of the military, the Beijing Academy of Military Medical Sciences is very interested in human cloning and stem cell research.

Chinese eugenics: Withholding technologies can be just as oppressive as imposing them. If the Catholic Church, for example, now managed to ban contraceptives or IVF, it would seem despotic even to many Catholics. Some would view a prohibition on prenatal testing and the subsequent abortions performed by women to avoid having a child with Down syndrome as even worse. Easy answers about who decides these matters will not appear, and much will hinge not on ethical values but on how countries balance individual desires with social needs. The problem, of course, is that gauging what will be good for society is inherently subjective, even when the consequences are clear. And they aren't. The vast majority of personal reproductive choices will have impacts on society that are distant and largely unknowable, and about which there would be little agreement anyway.

Try to have a boy: See Pomfret, 2000, p. A01; Farah, 1997; and Zeng, 1993, p. 297.

161 80 percent: See Macer, 1995. Similar numbers were reported in all other countries tested: New Zealand, Australia, Japan, India, Thailand, Israel, and Russia.

92 percent of counselors: See Wertz, 1999, pp. 144, 145, 149.

163 "Cloning a human baby": See http://pewforum.org/events/0503, the transcript of the Pew Forum on Religion and Public Life, "Human Cloning: Religious Perspectives," May 3, 2001.

164 Pent-up international demand: See Gardner, 1995. Some in public policy push for global solutions, but differences in culture and philosophy are probably too great for this to succeed. Moreover, global economic competition will work against this, because nations that push ahead with these technologies will likely reap the rewards of any medical spinoffs.

Gene expression errors: The most thorough research into the problems of genetic dysfunction in cloned animals has been done in mice: Humphreys (2001) measured gene expression for a half-dozen genes and reported unpredictable gene dysregulation in even apparently healthy cloned animals. In cattle, researchers have observed that much of the difficulty seems to arise because of abnormal formation of the placenta. Researchers are employing DNA chip technology to further study the process and make it more predictable. To give an idea of the current state of the art, using a normal unaltered embryo for IVF in cattle, about 50 percent result in pregnancies and most go to term. Using cloned embryos, nearly as many, perhaps 40 percent, lead to pregnancies at thirty-five days, but then about 90 percent of those are lost in the next two months. Of the 10 percent that reach term, perhaps two thirds seem healthy. (Estimates are from a conversation with Mark Westhusin, an expert on the faculty of Texas A&M.)

Public opposition will melt away: That genetically altered food (GAF), after years of nearly unnoticed use in the United States, has stirred such strong protest in Europe suggests that we may be in for a rough time. When "golden rice," brimming with vitamin A and promising to improve nutrition dramatically in underdeveloped nations, is rejected as unnatural, clearly we are on the cusp of a culture war, with genetic modification as the target. GAF has served as a focus for disparate political, religious, and environmental groups fearful of our genomic future, and the prospects of genetically altered humans will probably be even more threatening.

166 "Interventions in the human germline": See http://members.eunet.at/suepo/resen.htm.

166 Simply migrate elsewhere: Several American researchers working on human embryonic stem cells, for example, have tired of the obstacles to such research in the United States and have moved to Great Britain.

Testing the gender: At the 1996 Summer Olympics in Atlanta, 8 of 3,387 female athletes failed the genetic test. Most had a Y chromosome, but were feminized because of a resistance to testosterone. The genetic test was dropped before the 2000 Olympics. See Ferguson-Smith et al., 1992.

167 Fundamental right: *Eisenstadt v. Baird,* 405 US 438 (1972). See Robertson, 1996.

168 Guarantee them a deaf child: For a genetic mutation that would be predictive of deafness, see Li et al., 1998. Also see Kennedy, 1998.

Prenatal tests for sex-selection: See Nippert, 1999.

Policing is difficult: The obvious question is whether the performance itself is what matters or how the athlete achieved it. We can see an athletic event, categorize it, and define its boundaries. We can separate athletes into broad categories, though sometimes these categories change. The line between professional and amateur athletics became untenable amid the complexities of modern life and was dropped as a criterion for Olympic competitors. Specifying boundaries regarding the pharmaceutical history of athletes may become equally messy as drugs ever more closely match and mimic the natural compounds that inspire them.

170 "For a moment": See Turner, 1920, pp. 1–38. He first articulated this thesis in an influential paper he read at a meeting of the American Historical Association in Chicago, July 12, 1893.

171 "If you suddenly": See "Best Hope, Worst Fear: An Internet Multimedia Project on Human Germline Manipulation." http://research.mednet.ucla.edu/pmts/germline/bhwf.htm.

172 Test pilots: As late as 1970, 1 in 4 U.S. Navy test pilots died in accidents. We admire this risk-taking, just as we do valor in times of war. Wolfe, 1983.

173 "If G-d has built": Freundel, 2000.

174 Comforting to know: Even if the human form and character change radically, this new environment will likely be comfortable for our children, because they will grow up with it. People who imagine that we will somehow be able to shape the nature of future human life are deluding themselves. We can barely control today's world, much less the future one. If we can equip our kids with the best tools to make their own choices, we will be doing pretty well.

"I must altogether abandon": See de Santiallana, 1955, pp. 310–11.

175 Eons of time: What could be more awesome than that a large enough quantity of pure hydrogen exploded to form a universe and then gradually condensed into the intricate cellular structures of our flesh? Our evolution from primitive life seems pedestrian compared with the notion that the atoms in our bodies had to pass through the hearts of stars to reach us.

9. The Enhanced and the Unenhanced

177 Gene variants responsible for IQ: Predispositions for high IQ — and other cognitive attributes, for that matter — may be among the most complex of human traits. Though single-gene mutations have been implicated in various substantial diminutions of cognitive functioning, no specific genes have yet been found that account for even a few percentage points of variation in IQ among people with average and above-average scores. Current bioinformatic studies are too primitive — too small and crude — to identify combinations of alleles that together will raise IQ or to identify rare individual alleles that do. If researchers have so far failed to identify any strong single-gene contributions, this does not mean that combinations of genes will exert no strong effects, that there are no rare alleles that exert significant effects, or that no way will be found to manipulate relevant biochemical pathways. But if embryo selection for the main components of the heritable contribution to high IQ proves possible, this would still not usher in a genetic supermarket where parents could fill a shopping cart with multiple enhancements. Even without inherent biological tradeoffs among traits, finding the right gene combinations would require the screening of huge numbers of embryos.

nearly 120: This estimate assumes that IQ is only 50 percent heritable in typical environments (which is at the low end of current estimates; see, for example, Steen, 1999, pp. 113–35) and that only half of the total genetic variance is within individual families. The top embryo — selected on genetic criteria alone — would on average become an adult who tested at 118. By comparison, the top 1 percent of children have IQs of about 138.

178 Wish to differentiate: See Parens, 1998.

"The use of IGM": See Frankel and Chapman, 2000, p. 42.

180 Much easier to accomplish: Say IQ turned out to be very complex, shaped by twenty key genes and hundreds of others with minor influences. No doubt there would be many genes or small clusters of genes that when mutated cause nutritional problems or otherwise disrupt normal brain development, resulting in diminished intelligence. PKU (phenylketonuria) is one such condition described by Paul (1998, p. 178), and it is relatively straightforward to repair. But the task of improving on a rare combination of genes that contribute to genius would be far more difficult and demand greater caution about unseen tradeoffs.

181 It is inconceivable: See Scott and Fuller, 1974, pp. 403, 411.

182 Unwitting pilot project: See Vila et al., 1997, and Wayne, 1993. Although the oldest archaeological evidence of the association between dogs and people dates back only about 14,000 years, mitochondrial sequencing suggests that dogs branched off from gray wolves around 135,000 years ago. (Gray wolves had branched off from coyotes and foxes 5 to 10 million years earlier.) It is

possible that dogs did not diverge anatomically from wolves until humans began to inhabit agricultural centers and impose stronger selective pressures. Few dog breeds can be traced back more than a few thousand years, and most have appeared only in the past few centuries. This may seem a short time, but a single breeder's hand can direct their evolution for thirty or more dog generations, whereas our evolution, with so few generations encompassed by any single human lifetime, must be directed by larger social and cultural forces.

In a forty-year domestication experiment on foxes, Russian breeders, using modern methods, selected for a single trait: tameness. This, they believe, is the common trait that all human domestication has selected, and it produced aspects of doglike morphology, coat color variability, size changes, tail changes, and such. See Trut, 1999. www.blarg.net/~critter/dogfamily/ancientdog_3.htm.

182 Still a single species: When researchers compared a hundred different genetic markers from ninety-six dog breeds, they couldn't distinguish one breed from another. This means that the differences are much more fine-grained, encompassing relatively small numbers of genes. These differences may give us great insight into the potential for directing human evolution. Certainly, such efforts have been relatively easy with dogs and foxes. (Personal communication, Jasper Rine, University of California at Berkeley.)
Biological forms that persist: See Raup, 1991, p. 108.

183 Cultural rather than biological: See Blackmore, 1999.

184 Development of writing: See Stock, 1993, p. 85.

185 Birth in a hospital: See Wertz and Wertz, 1989.

186 Alter specific genes: See Capecchi, 2000.
$6,000 a baby: The cost of a viable pregnancy by IVF for a thirty-year-old woman without serious fertility problems is now about $12,000. Assuming savings from new technology, IVF that is routine, massive numbers of procedures, and automated GCT, the lower figure is not outlandish; it costs about the same amount to order a strain of knock-out mice from a laboratory.

187 Scoring above 700: See Cook and Frank, 1991.

189 Similarity of all life: These figures are not precise, since we don't even know the exact number of human genes, but they are in the ballpark. For a good general discussion of the human genome, see the *Nova* interview with Eric Landers, the director of the Center for Genome Research at the Whitehead Institute: www.pbs.org/wgbh/nova/genome/deco_lander.html.
The differences between us: We are a young species, all coming from a population of a few tens of thousands of people that existed 100,000 years ago in Africa. Because this founder population was so small and recent, the differences between two humans are perhaps a quarter as great as those between two chimpanzees.

190 Around 10 to 20 percent: This would include the coding regions of genes, which determine the sequences of the protein or proteins each gene specifies, and the non-coding regions, which regulate the expression of the genes. Together these account for perhaps 5 percent of the human genome.

191 Distaste for GCT: While some people fear that GCT will be narrowly held and therefore lead to a genetic elite, others fear that the multitudes will have access to GCT and make pathological or at least unwise choices. And some people fear both. For example, Sheldon Krimsky, a professor at Tufts University, writes, "The availability of eugenic techniques in reproduction to a minority of affluent people will support the 'geneticization' of a society, enabling an aristocracy with so-called proper genes to use it to their class advantage." This suggests that such choices would be of value. But elsewhere in the same essay, he suggests that "offering people the opportunity to choose the phenotype of a child will result in psychosocial pathologies, including deeper class and racial divisions within society" (Krimsky, 2000). Apparently, only the elite would have the sense to use the technology to their benefit.

Effort to block: The so-called War on Drugs provides a cautionary tale for GCT. This war has filled our jails, corrupted swaths of law enforcement and government, made criminals out of many otherwise law-abiding citizens, and funneled enormous resources into the hands of criminals. But many supporters of it contend that legalization would be far worse, because it would bring about greater use of drugs and destroy countless lives. Duke and Gross (1994) presents an excellent argument against the War on Drugs.

192 Split into separate breeds: Were enhancements based on changes to our 46 existing chromosomes rather than on replaceable artificial chromosomes, each generation would have to build on the changes made by the previous one, so family lineages would progressively accumulate benefits and diverge from one another. In essence, everyone's genetic choices about their offspring would be circumscribed by the past decisions of their ancestors.

193 Will narrow human diversity: The screening out of genetic vulnerabilities would no doubt also result in fewer individuals whose afflictions bring about uncommon achievements — the manic depressive who writes great literature, the physically impaired artist who creates great works of art. In any event, in the future many such afflictions will be increasingly blunted by improved medical treatment.

195 Self-directed evolution: Campbell (1995) explores the concept of self-directed evolution at length. He calls it "regenerative evolution" and believes that small groups of future humans, highly committed to self-evolution, will outstrip humanity as a whole, leaving it far behind. In his view, how humanity responds to GCT is irrelevant, because small founder groups will be at the heart of any future evolutionary change.

197 Self-domestication: This process mirrors our domestication of dogs, cats,

livestock, and crops. By selecting for those qualities that bring these species into our lives, we have transformed them, and we have transformed ourselves through a similar process of self-selection. Our transformation has been primarily cultural, but it has almost certainly had a biological component: selection for the traits that allow survival in the altered world we have been creating.

197 Shackleton expedition: See Worsley, 1931.

198 In the 1920s: Kevles (1995) provides an in-depth look of the origins of the eugenics movement. The mainstream movement was largely oriented toward voluntarism and relied on education, contraception, and moral injunction. Eugenic ideas were so commonly accepted that a 1937 poll in *Fortune* magazine showed 63 percent of Americans endorsing the compulsory sterilization of habitual criminals and 66 percent in favor of sterilizing "mental defectives." In 1939, Hermann Muller, who later won the Nobel Prize and helped found the "genius" sperm bank in San Diego, published a "Genetics Manifesto" with twenty-two other scientists asserting that it should be "an honor and a privilege, if not a duty, for a mother, married or unmarried, or for a couple, to have the best children possible, both in upbringing and genetic endowment."

199 Global policies: See Macer, 2000.

Appendix 1

206 Basic constitutional right: See Robertson, 1994, ch. 2.
Human cloning: For a discussion of potential psychological damage to a clone, see Pence, 1998.

208 Fetal screening: See www.chi.swan.ac.uk/hosted/maternity/nevillhl.htm. An added dilemma for mothers is that such testing is itself dangerous. About 1 percent of the time, amniocentesis leads to miscarriage, and generally, the lost child would have been normal.

209 Legal and legislative tools: Dworkin (1996) takes a thoughtful look at the dangers of ill-considered legal and legislative responses to the challenges of biotechnology and medicine.

Appendix 2

210 The questions are adapted from Stock (2002), a collection of value-laden questions about the personal challenges and dilemmas that will attend our rapid technological advance. The book touches on coming biological change, artificial intelligence, pervasive telecommunications, space travel, electronic monitoring, and more. To join a global discussion of the issues raised here, you may participate in a forum for this purpose: www.bookofquestions.com/Redesigninghumans.

Bibliography

Alexander, B. 2000. "Don't Die, Stay Pretty: Introducing the Ultrahuman Makeover." *Wired,* January.

Alexander, B. 2001. "(You)^{2}" *Wired,* February.

American Society for Pharmacology and Experimental Therapeutics. 1999. *Public Affairs News Brief,* no. 14 (August). www.faseb.org/aspet/PABNo14_99.htm.

Ames, B. 1998. "Micronutrients Prevent Cancer and Delay Aging." *Toxicology Letters* 102: 5–18.

Ash, E. 1927. *Dogs: Their History and Development.* 2 vols. London: Ernest Benn Limited.

Atkinson, M., and E. Leiter. 1999. "The NOD Mouse Model of Type 1 Diabetes: As Good as It Gets?" *Nature Medicine* 5: 601–4.

Baharloo, S., et al. 1998. "Absolute Pitch: An Approach for Identification of Genetic and Nongenetic Components." *American Journal of Human Genetics* 62: 224–31.

Baharloo, S., et al. 2000. "Familial Aggregation of Absolute Pitch." *American Journal of Human Genetics* 67: 755–58.

Bailey, R. 1998. "Send in the Clones." *Reason,* June.

——. 2001. "Rage Against the Machines: Witnessing the Birth of the Neo-Luddite Movement." *Reason,* July, 26–35.

Baird, P., et al. 1988. "Genetic Disorders in Children and Young Adults: A Population Study." *American Journal of Human Genetics* 42: 677–93.

Bardo, S. 1998. "Braveheart, Babe, and the Contemporary Body." In *Enhancing Human Traits.* Edited by Erik Parens. Washington, D.C.: Georgetown University Press.

Barritt J., et al. 2001. "Mitochondria in Human Offspring Derived from Ooplasmic Transplantation." *Human Reproduction* 16: 513–16.

Benjamin, J., et al. 1996. "Population and Familial Association Between the D4

Dopamine Receptor Gene and Measures of Novelty Seeking." *Nature Genetics* 12: 81–84.

Bertelsen, A., et al. 1977. "A Danish Twin Study of Manic-Depressive Disorders." *Archives of General Psychiatry* 130: 330–51.

Biasolo, M. A., et al. 1996. "A New Antisense tRNA Construct for the Genetic Treatment of Human Immunodeficiency Virus Type 1 Infection." *Journal of Virology* 70: 2154–61.

Blackmore, S. 1999. *The Meme Machine.* New York: Oxford University Press.

Bohmann, M., et al. 1982. "Predisposition to Petty Criminality in Swedish Adoptees: Genetic and Environmental Heterogeneity." *Archives of General Psychiatry* 39: 1233–41.

Bonnicksen, A. 1994. "National and International Approaches to Human Germline Gene Therapy." *Politics and the Life Sciences* 13 (1), 38–49.

Bouchard, T. 1994. "Genes, Environment, and Personality." *Science* 264: 960–62.

Bouchard T., et al. 1981. "The Minnesota Study of Twins Reared Apart: Project Description and Sample Results in the Developmental Domain." *Progress in Clinical and Biological Research* 69B: 227–33.

——. 1990. "Sources of Human Psychological Differences: The Minnesota Study of Twins Reared Apart." *Science* 250: 223–28.

Brody, J. 1997. "Obesity Drugs: Weighing the Risks to Health Against the Small Victories." *New York Times,* September 3.

Browning, J., et al. 1997. "Mice Transgenic for Human CD4 and CCR5 Are Susceptible to HIV Infection." *Proceedings of the National Academy of Sciences* 94: 14637–41.

Brum, G., et al. 1994. *Biology: Exploring Life, Cell Biology, and Genetics.* New York: John Wiley & Sons.

Bunting, M., et al. 1999. "Targeting Genes for Self-Excision in the Germline." *Genes and Development* 13: 1524–28.

Butler, S. 1872. *Erewhon.* New York: Viking Penguin (1970).

Campbell, J. 1995. "The Moral Imperative of Our Future Evolution." In *Evolution and Human Values,* edited by R. Wesson and P. Williams. New York: Rodopi.

Campbell, J., and G. Stock. 2000. "A Vision for Practical Human Germline Engineering." In *Engineering the Human Germline: An Exploration of the Science and Ethics of Altering the Genes We Pass to Our Children,* edited by G. Stock and J. Campbell, 9–16. New York: Oxford University Press.

Candy C., et al. 2000. "Restoration of a Normal Reproductive Lifespan after Grafting of Cryopreserved Mouse Ovaries." *Human Reproduction* 15: 1300–1304.

Capecchi, M. 1989. "The New Mouse Genetics: Altering the Genome by Gene Targeting." *Trends in Genetics* 5: 70–76.

——. 1989. "Altering the Genome by Homologous Recombination." *Science* 244: 1288–92.

——. 2000. "Human Germline Gene Therapy: How and Why." In *Engineering the Human Germline: An Exploration of the Science and Ethics of Altering the Genes We Pass to Our Children,* edited by G. Stock and J. Campbell, 31–42. New York: Oxford University Press.

Cavazzana-Calvo, M., et al. 2000. "Gene Therapy of Severe Combined Immunodeficiency (SCID) — X1 Disease." *Science* 288: 669–72.

Chan, W., et al. 2001. "Transgenic Monkeys Produced by Retroviral Gene Transfer into Mature Oocytes." *Science* 291: 309–12.

Children's Bureau. 1950. *Changes in Infant, Childhood, and Maternal Mortality over the Decade of 1939–1948: A Graphic Analysis.* Washington, D.C.: Children's Bureau, Social Security Administration.

Chislenko, A. 1995. "Legacy Systems and Functional Cyborgization of Humans." www.lucifer.com/~sasha/articles/Cyborgs.html.

Choo, A. 2001. "Engineering Human Chromosomes for Gene Therapy Studies." *Trends in Molecular Medicine* 7: 235–37.

Christopherson, K., et al. 1992. "Ecdysteroid-Dependent Regulation of Genes in Mammalian Cells by a Drosophila Ecdysone Receptor and Chimeric Transactivators." *Proceedings of the National Academy of Sciences* 89: 6314–18.

Cloninger, C., et al. 1996. "Type I and Type II Alcoholism: An Update." *Alcohol Health and Research World* 20: 18–24.

Coghlan, H. 1999. "We Have the Power: A Safer Way of Altering Genes Will Make Engineering Humans More Tempting than Ever." *New Scientist,* October 23.

Colledge, W., et al. 1995. "Generation and Characterization of a Delta F508 Cystic Fibrosis Mouse Model." *Nature Genetics* 10: 445–52.

Cook, P., and R. Frank. 1991. "The Growing Concentration of Top Students at Elite Schools." In *The Economics of Higher Education,* edited by C. Clotfelter and M. Rothschild. Chicago: NBER-University of Chicago Press.

Cook-Deegan, R. 1994. *The Gene Wars: Science, Politics, and the Human Genome.* New York: W. W. Norton.

Council on Scientific Affairs, American Medical Association. 1992. "Induced Termination of Pregnancy before and after Roe v. Wade: Trends in the Morbidity and Mortality of Women." *Journal of the American Medical Association* 268 (22): 3231–39.

Dahlburg, J. 2000. "French Cycling Takes a Spill at Doping Trial." *Los Angeles Times,* October 28, A2.

Daly, M., and M. Wilson. 1988. *Homicide.* New York: Aldine de Gruyter.

Dam, J., and U. Wihlborg. 2000. "At 52, Cheryl Tiegs Awaits Twins." *People,* June 12.

Deborah, O., et al. 2000. "Generation of Transgenic Mice and Germline Transmission of a Mammalian Artificial Chromosome Introduced into Embryos by Pronuclear Microinjection." *Chromosome Research* 8: 183–91.

de Grey, A., et al. 2001. "Time to Talk SENS: Critiquing the Immutability of Human Aging." In *Increasing Healthy Life Span: Conventional Measures and Slowing the Innate Aging Process.* Proceedings of the Ninth Congress of the International Association of Biomedical Gerontology. *Annals of the New York Academy of Sciences,* edited by D. Harman (in press).

DeLisi, C. 1988. "The Human Genome Project." *American Scientist* 76: 488.

de Santillana, G. 1955. *The Crime of Galileo.* Chicago: University of Chicago Press.

de Waal, F. 1998. *Chimpanzee Politics: Power and Sex among Apes.* Baltimore: Johns Hopkins University Press.

DeWitt, P. 1993. "In Pursuit of Pregnancy." *American Demographics,* May.

Diamond, J. 1993. *The Third Chimpanzee: Evolution and Future of the Human Animal.* New York: Harper Perennial.

Dickemann, M. 1979. "Female Infanticide and Reproductive Strategies of Stratified Human Societies." In *Evolutionary Biology and Human Social Behavior,* edited by N. Chagnon, 321–67. North Scituate, Mass.: Duxbury Press.

Dobelle, W. 2000. "Artificial Vision for the Blind by Connecting a Television Camera to the Visual Cortex." *ASAIO Journal* (American Society for Artificial Internal Organs), January. www.artificialvision.com/vision/asaio1.html.

Dove, A. 1999. "Research News: Germline Transmission of Artificial Chromosome." *Nature Biotechnology* 17 (12): 1149.

Duan, L., et al. 1997. "Intracellular Inhibition of HIV-1 Replication Using a Dual Protein and RNA-Based Strategy." *Gene Therapy* 4: 533–43.

Duke, S., and A. Gross. 1994. *America's Longest War: Rethinking Our Tragic Crusade Against Drugs.* New York: Putnam.

Dulbecco, R. 1986. "A Turning Point in Cancer Research: Sequencing the Human Genome." *Science* 231: 1055.

Dworkin, R. 1996. *Limits: The Role of the Law in Bioethical Decision Making.* Bloomington: Indiana University Press.

Edirisinghe W., et al. 1997. "Birth from Cryopreserved Embryos Following In-Vitro Maturation of Oocytes and Intracytoplasmic Sperm Injection." *Human Reproduction* 12: 1056–58.

Edwards, A. 1987. "Male Violence in Feminist Theory: An Analysis of Sex/Gender Violence and Male Dominance." In *Women, Violence, and Social Control,* edited by J. Hanmer and M. Maynard. London: Macmillan.

Efron, S. 2001. "Baby Bust Has Japan Fearing for Its Future." *Los Angeles Times,* June 24, A1.

———. 2001. "An Aging Population Creates a 'Nursing Hell' for Many Women." *Los Angeles Times,* June 25, A1.

Entine, J. 2000. *Taboo: Why Black Athletes Dominate Sports and Why We're Afraid to Talk about It.* New York: Public Affairs.

Epstein, R., et al. 1995. "Dopamine D4 Receptor (D4DR) Exon III Polymorphism Associated with the Human Personality Trait of Novelty Seeking." *Nature Genetics* 12: 78–80.

Farah, J. 1997. "Cover-up of China's Gender-cide." World Net Daily Archive, September 29.

Feldman, M. 1998. "Move Over, Mother Nature: Making Babies with Mother Science." *Minnesota Medicine Journal* 81 (October).

Fenwick, L. 1997. *Private Choices, Public Consequences.* New York: Dutton.

Ferber, R. 1986. *Solve Your Child's Sleep Problems.* New York: Simon & Schuster.

Ferguson-Smith, M., et al. 1992. "Olympic Row over Sex Testing." *Nature* 355: 10.

Finch, C., and T. Kirkwood. 1999. *Chance, Development, and Aging.* New York: Oxford University Press.

Finkel, E. 2001. "Engineered Mouse Virus Spurs Bioweapon Fears." *Science* 291: 585.

Fisher, F., and A. Sommerville. 1998. "To Everything There Is a Season? Are There Medical Grounds for Refusing Fertility Treatment to Older Women?" In *The Future of Human Reproduction: Ethics, Choice, and Regulation,* edited by J. Harris and S. Holm. New York: Oxford University Press.

Fletcher, J. 1994. "Germ-Line Gene Therapy: The Costs of Premature Ultimates." *Politics and the Life Sciences,* August, 225–27.

Fletcher, R. 1991. "Intelligence, Equality, Character, and Education." *Intelligence* 15: 139–49.

Frankel, M., and A. Chapman. 2000. *Human Inheritable Genetic Modifications: Assessing Scientific, Ethical, Religious, and Policy Issues.* Washington, D.C.: AAAS Publication Services. www.aaas.org/spp/dspp/sfrl/germline/main.htm.

Frankenfield, G. 2000. "The Nash Family: Breaking New Ground in Medicine." WebMD.com, October 4.

Freundel, B. 2000. "Gene Modification Technology." In *Engineering the Human Germline: An Exploration of the Science and Ethics of Altering the Genes We Pass to Our Children,* edited by G. Stock and J. Campbell, 121. New York: Oxford University Press.

Galloway, S., et al. 2000. "Mutations in an Oocyte-Derived Growth Factor Gene (BMP15) Cause Increased Ovulation Rate and Infertility in a Dosage-Sensitive Manner." *Nature Genetics* 25 (3): 279–83.

Gao, X., et al. 1999. "Advanced Transgenic and Gene-Targeting Approaches." *Neurochemical Research* 24: 1181–88.

Gardner, W. 1995. "Can Human Genetic Enhancement Be Prohibited?" *Journal of Medicine and Philosophy* 20: 65–84.

Gelsinger, P. 2000. Testimony before the Subcommittee on Public Health of the Health, Education, Labor, and Pension Committee of the U.S. Senate. February 2.

Gems, D. 2002. "Treatment, Enhancement and the Retardation of Ageing: Is More Life Always Better?" In preparation.

Genillard, A. 1993. "Genetic Research Curbs to Be Eased." *Financial Times,* May 28.

Gibson, W. 1986. "Johnny Mnemonic." In *Burning Chrome.* New York: Arbor House.

Golic, K., and M. Golic. 1996. "Engineering the Drosophila Genome: Chromosome Rearrangements by Design." *Genetics* 144: 1693–1711.

Golic, K., and S. Lindquist. 1989. "The FLP Recombinase of Yeast Catalyzes Site-Specific Recombination in the Drosophila Genome." *Cell* 59: 499–509.

Gosden, R. G., et al. 1993. "The Biology and Technology of Follicular Oocyte Development In Vitro." *Reproductive Medicine Review* 2: 129–52.

Gottesman, I. 1993. "Origins of Schizophrenia: Past as Prologue." *Nature, Nurture, and Psychology,* edited by R. Plomin and G. McClearn. Washington, D.C.: American Psychological Association.

Gould, S. J. 1994. "Curveball." *The New Yorker,* November 28, 139–49.

Grady, D. 1999. "Doctor Devoted to Treating Kids with Inherited Liver Disorder." *New York Times,* July 10, P2.

Gregersen, P., et al. 1999. "Absolute Pitch: Prevalence, Ethnic Variation, and Estimation of the Genetic Component." Letter. *American Journal of Human Genetics* 65: 911–13.

Grody, W. 1999. "Cystic Fibrosis: Molecular Diagnosis, Population Screening, and Public Policy." *Archives of Pathology and Laboratory Medicine* 123: 1041–46.

Hamer, D., and P. Copeland. 1999. *Living with Our Genes.* New York: Anchor Books.

Hamer, D., et al. 1993. "A Linkage Between DNA Markers on the X Chromosome and Male Sexual Orientation." *Science* 261: 321–27.

Hammond, C., et al. 2001. "Genes and Environment in Refractive Error: The Twin Eye Study." *Investigative Ophthalmology and Visual Science* 42: 1232–36.

Handyside, A., et al. 1990. "Pregnancies from Biopsied Human Preimplantation Embryos Sexed by Y-Specific DNA Amplification." *Nature* 244: 768–70.

Handyside, A. H., et al. 1992. "Birth of a Normal Girl after In Vitro Fertilization and Preimplantation Diagnostic Testing for Cystic Fibrosis." *New England Journal of Medicine* 327: 905–9.

Harrington, J., et al. 1997. "Formation of De Novo Centromeres and Construction of First-Generation Human Artificial Microchromosomes." *Nature Genetics* 15: 345–55.

Harvey, D. M., and C. T. Caskey. 1998. "Inducible Control of Gene Expression: Prospects for Gene Therapy." *Current Opinion in Chemistry and Biology* 2: 512–18.

Hatch, E., et al. 1998. Cancer Risk in Women Exposed to Diethylstilbestrol *In Utero*." *Journal of the American Medical Association* 280: 630–34.

Hattori, M., et al. 2000. "The DNA Sequence of Chromosome 21." *Nature* 405: 311–19.

Herrnstein, R., and C. Murray. 1994. *The Bell Curve: Intelligence and Class Structure in American Life*. New York: Free Press.

Herskind, A., et al. 1996. "The Heritability of Human Longevity: A Population-Based Study of 2872 Danish Twin Pairs Born 1870–1900." *Human Genetics* 97(3): 319–23.

Horgan, J. 1993. "Genes and Crime: A U.S. Plan to Reduce Violence Rekindles an Old Controversy." *Scientific American* 268 (February): 24.

Hubbard, R. 2000. "Germline Manipulation." In *Engineering the Human Germline: An Exploration of the Science and Ethics of Altering the Genes We Pass to Our Children*, edited by G. Stock and J. Campbell, 109–11. New York: Oxford University Press.

Humpherys, D., et al. 2001. "Epigenetic Instability in ES Cells and Cloned Mice." *Science* 293: 95–97.

International Human Genome Sequencing Consortium, The. 2001. "Initial Sequencing and Analysis of the Human Genome." *Nature* 409: 860.

Isachenko, E., et al. 2000. "Progress, Problems and Perspectives in Cryopreservation of Human Oocytes, Embryos, and Ovarian Tissue." *Fertimagazine*, September.

Jaenisch, R., and I. Wilmut. 2001. "Developmental Biology: Don't Clone Humans." *Science* 291: 2552.

Jensen, A. 1998. *The G Factor: The Science of Mental Ability*. New York: Praeger.

Jensen, A., and E. Munro. 1970. "Reaction Time, Movement Time, and Intelligence." *Intelligence* 3: 121–26.

Jones, D. 2000. "On Moore's Law and Fishing: Gordon Moore Speaks Out." *U.S. News & World Report*, July 10.

Joynson, R. 1991. *The Burt Affair*. London: Routledge.

Kan, H., et al. 2000. "Genetic Testing — Present and Future." *Science* 289: 1890–92.

Kass, L. 2001. "Why We Should Ban Cloning Now: Preventing a Brave New World." *New Republic*, May 21, 30–39.

Kast, E. 2000. Testimony on behalf of the Cystic Fibrosis Foundation before the Subcommittee on Public Health of the Health, Education, Labor, and Pension Committee of the U.S. Senate, February 2.

Kay, M., et al. 2000. "Evidence for Gene Transfer and Expression of Factor IX in Haemophilia B Patients Treated with an AAV Vector." *Nature Genetics* 24: 257–61.

Keating, S. 2001. "Sports Doping Body Worried about Gene Engineering." Associated Press, May 16.

Kennedy, M. 1998. "Ethics of Genetic Testing." *Wisconsin Medical Journal,* April. www.wismed.org/wmj/98–04/wmj498-kenn3.htm.

Kenyon, W. 1996. "Ponce d'Elegans: Genetic Quest for the Fountain of Youth." *Cell* 84: 501–4.

Kevles, D. 1995. *In the Name of Eugenics: Genetics and the Uses of Human Heredity.* Cambridge: Harvard University Press.

King, R., and W. Stansfield. 1997. *A Dictionary of Genetics,* 5th ed. New York: Oxford University Press.

Kitamura, K. 1998. "Transgene Regulation by the Tetracycline-Controlled Transactivation System." *Experimental Nephrology* 6: 576–80.

Kmietowicz, Z. 2001. "Genes Implicated in Vision Problems." Associated Press newswire, May 25.

Koenig, R. 2001. "Sardinia's Mysterious Male Methuselahs." *Science* 291: 2074–76.

Kolata, G. 1998. "Scientists Brace for Changes in Path of Human Evolution." *New York Times,* March 21, A1.

———. 1999. "Pushing Limits of the Human Lifespan." *New York Times,* March 9, D1.

Kornblut, A. 2001. "Stem Cell Debate Forces Many to Reconsider Their Ethical Positions." *Boston Globe,* July 29, A1.

Kren B., et al. 1999. "Correction of the UDP-Glucuronosyltransferase Gene Defect in the Gunn Rat Model of Crigler-Najjar Syndrome Type I with a Chimeric Oligonucleotide." *Proceedings of the National Academy of Sciences* 96: 10349–54.

Krimsky, S. 2000. "The Psychosocial Limits on Human Germline Modification." In *Engineering the Human Germline: An Exploration of the Science and Ethics of Altering the Genes We Pass to Our Children,* edited by G. Stock and J. Campbell, 104–7. New York: Oxford University Press.

Kristof, N. 1991. "Asia, Vanishing Point for as Many as 100 Million Women." *International Herald Tribune,* November 6, 1.

Kuhn, R. 2000. *Closer to Truth: Challenging Current Belief.* New York: McGraw-Hill.

Kuleshova, L., et al. 1999. "Birth Following Vitrification of a Small Number of Human Oocytes: Case Report." *Human Reproduction* 14: 3077–79.

Kurzweil, R. 1999. *The Age of Spiritual Machines.* New York: Viking Penguin, 279, 280.

Lander, E. 2000. "After Deciphering the Map, the Next Task Is a Guidebook for the Human Genome." *New York Times,* September 12.

Lanier, J. 2000. "Mindless Thought Experiments: A Critique of Machine Intelligence." www.advanced.org/jaron/aichapter.html.

———. 2000. "A Tale of Two Terrors." *CIO Magazine,* July 1. www.cio.com/archive/070100_diff.html.

Lanza, R., et al. 2000. "Cloning of an Endangered Species (*Bos gaurus*) Using Interspecies Nuclear Transfer." *Cloning* 2 (2): 79–90.

Leavitt, M. C., et al. 1994. "Transfer of an Anti-HIV-1 Ribozyme Gene into Primary Human Lymphocytes." *Human Gene Therapy* 5: 1115–20.

Lederberg, J. 2000. "Pathways of Discovery: Infectious History." *Science* 288: 293.

Lee, C., et al. 2000. "Gene-Expression Profile of the Ageing Brain in Mice." *Nature Genetics* 25: 294–97.

Lee, C.-K., et al. 1999. "Gene Expression Profile of Aging and Its Retardation by Caloric Restriction." *Science* 285: 1390–93.

Lesch, K., et al. 1996. "Association of Anxiety-Related Traits with a Polymorphism in the Serotonin Transporter Gene Regulatory Region." *Science* 274: 1527–31.

Lewis, R. 1998. "Mammalian Cloning Milestone: Mice from Mice from Mice." *Scientist* 12 (16): 1, 7.

Li, C., et al. 1998. "A Mutation in PDS Causes Non-Syndromic Recessive Deafness." *Nature Genetics* 18: 215–17.

Loehlin, J. C., et al. 1989. "Modeling IQ Change: Evidence from the Texas Adoption Project." *Child Development* 60: 993, 1004.

Macer, D. 2000. "Universal Bioethics for the Human Germline." In *Engineering the Human Germline: An Exploration of the Science and Ethics of Altering the Genes We Pass to Our Children,* edited by G. Stock and J. Campbell, 139–141. New York: Oxford University Press.

Macer, D., et al. 1995. "International Perceptions and Approval of Gene Therapy." *Human Gene Therapy* 6: 791–803.

Mark, E., et al. 1997. "Brief Report: Fatal Pulmonary Hypertension Associated with Short-Term Use of Fenfluramine and Phentermine." *New England Journal of Medicine* 337 (9): 581–89.

Marshall, E. 2001. "Company Plans to Bank Human DNA Profiles." *Science* 291: 575.

Marx, J. 1994. "Mouse Model Found for ALS." *Science* 264 (5166): 1663–64.

McCartney, K., et al. 1990. "Growing Up and Growing Apart: A Developmental Meta-Analysis of Twin Studies." *Psychological Bulletin* 107: 226–37.

McCue, M., et al. 1993. "Behavioral Genetics of Cognitive Ability: A Life-Span Perspective." In *Nature, Nurture, and Psychology,* edited by R. Plomin and G. McClearn. Washington, D.C.: American Psychological Association.

McGleenan, T. 1995. "Human Gene Therapy and Slippery Slope Arguments." *Journal of Medical Ethics* 21 (6): 350–55.

Mestel, R. 2000. "Study Ties Most Cancer to Lifestyle, Not Genetics." *Los Angeles Times,* July 13, A1.

——. 2001. "Fully Internal Heart Device Installed." *Los Angeles Times,* July 4, A1.

Misteli, T. 2001. "Protein Dynamics: Implications for Nuclear Architecture and Gene Expression." *Science* 291: 843–47.

Montgomery, H., et al. 1998. "Human Gene for Physical Performance." *Nature* 393 (6682): 221–22.

Moravec, H. 1988. *Mind Children: The Future of Robot and Human Intelligence.* Cambridge: Harvard University Press.

More, M. 1999. "A Letter to Mother Nature." Keynote address, EXTRO-4 conference, Berkeley, California, August. www.maxmore.com/mother.htm.

Murphy, K., and R. Topel. 1999. "Medical Research: What's It Worth?" www.milken-inst.org/mod35/mir5_22_medresearch.pdf.

Napoli, M. 2000. "Preventing Medical Errors: A Call to Action." *Healthfacts,* January.

Narayanan, K., et al. 1999. "Efficient and Precise Engineering of a 200kb β-Globin Human/Bacterial Artificial Chromosome in *E. coli* DH10B Using an Inducible Homologous Recombination System." *Gene Therapy* 6: 442–47.

Neisser, U., et al. 1996. "Intelligence: Knowns and Unknowns." Report of a task force established by the Board of Scientific Affairs of the American Psychological Association. *American Psychologist,* February, 77–101.

Nippert, N. 1999. "Key Policy Issues in the Provision of Genetic Services in Germany." In *Chinese Scientists and Responsibility,* edited by O. Döring, 118–40. Hamburg: Mitteilungen des Instituts für Asienkunde.

Nordhaus, W. 1999. "The Health of Nations: The Contribution of Improved Health to Living Standards." www.econ.yale.edu/~nordhaus/homepage/health%20nnp%20111799.pdf.

O'Brien, C. 1996. "China Urged to Delay 'Eugenics' Law." *Nature* 383 (6597): 204.

Olshansky, S., and B. Carnes. 2000. *The Quest for Immortality: Science at the Frontiers of Aging.* New York: W. W. Norton.

Olshansky, S., et al. 2001. "Policy Forum: Prospects for Human Longevity." *Science* 291: 1491–92.

Olson, S. 2001. "The Genetic Archaeology of Race." *Atlantic Monthly,* April, 80.

Palermo, G., et al. 1992. "Pregnancies after Intracytoplasmic Injection of Single Spermatozoon into an Oocyte." *Lancet,* 340: 17–18.

Paleyanda, R., et al. 1997. "Transgenic Pigs Produce Functional Human Factor VIII in Milk as Well." *Nature Biotechnology* 15: 971–75.

Parens, E. 2000. "Justice and the Germline." In *Engineering the Human Germline: An Exploration of the Science and Ethics of Altering the Genes We Pass to Our Children,* edited by G. Stock and J. Campbell, 123. New York: Oxford University Press.

Parens, E., and E. Juengst. 2001. "Inadvertently Crossing the Germline Barrier." Editorial. *Science* 292: 397.

Parens, E., ed. 1998. *Enhancing Human Traits: Ethical and Social Implications.* Washington, D.C.: Georgetown University Press.

Pascal, B. 1670. *Pensées,* translated by A. Krailsheimer. Baltimore: Penguin (1966).

Patai, D. 2000. *Heterophobia: Sexual Harassment and the Future of Feminism.* Lanham, Md.: Rowman & Littlefield.

Patai, D., and N. Koertge. 1999. *Professing Feminism: Cautionary Tales from the Strange World of Women's Studies.* New York: Basic Books.

Patel, R. n.d. "The Practice of Sex Selection Abortion in India: May You Be the Mother of a Hundred Sons." www.ibiblio.org/ucis/pubs/Carolina_Papers/abortion.pdf.

Paul, D. 1998. *The Politics of Heredity: Essays on Eugenics, Biomedicine, and the Nature-Nurture Debate.* Albany: State University of New York Press.

Pence, G. 1998. *Who's Afraid of Human Cloning?* Lanham, Md.: Rowan & Littlefield.

Pickrell, J. 2001. "Experts Assail Plan to Help Childless Couples." *Science* 291: 2061–63.

Pomfret, J. 2000. "China Losing 'War' on Births: Uneven Enforcement Undermines One-Child Policy." *Washington Post Foreign Service,* May 3, A1.

Prather, R. 2000. "Pigs Is Pigs." *Science* 289: 1886–87.

Pugh, T., et al. 1999. "Controlling Caloric Intake: Protocols for Rodents and Rhesus Monkeys." *Neurobiology of Aging* 20: 157–65.

Rao, R. 1986. "Move to Stop Sex-Test Abortion." *Nature* 324: 202.

Raup, D. 1991. *Extinction: Bad Genes or Bad Luck?* New York: W. W. Norton.

Regis, E. 1990. *Great Mambo Chicken and The Transhuman Condition: Science Slightly Over the Edge.* Reading, Mass.: Addison-Wesley.

Rice, G., et al. 1999. "Male Homosexuality: Absence of Linkage to Microsatellite Markers at Xq28." *Science* 284: 665–67.

Richter, P. 2000. "Pilotless Plane Pushes Envelope for U.S. Defense." *Los Angeles Times,* May 14, A1.

Ridley, M. 1993. *The Red Queen: Sex and the Evolution of Human Nature.* New York: Penguin.

———. 1999. *Genome.* New York: HarperCollins.

Rifkin, J., and T. Howard. 1977. *Who Shall Play God?* New York: Dell.

Ritter, M. 2000. "Scientists Find Fat Gene." Associated Press newswire, March 27.

Roberts, L. 2001. "Controversial from the Start." *Science* 291: 1182–88.

Robertson, J. 1994. *Children of Choice: Freedom and the New Reproductive Technologies.* Princeton, N.J.: Princeton University Press.

———. 1996. "Genetic Selection of Offspring Characteristics." *Boston University Law Review* 76: 421–82.

Rosenberg, L., and A. Schechter. 2000. "Gene Therapist, Heal Thyself." *Science* 287: 1751.

Rotman, D. 2001. "Molecular Computing." *MIT Technology Review,* May–June, 52–59.

Rutter, M., et al. 1993. "Autism: Syndrome Definition and Possible Genetic Mechanisms." In *Nature, Nurture, and Psychology,* edited by R. Plomin and

G. McClearn, 269–84. Washington, D.C.: American Psychological Association.

Sapolsky, R. 2000. "Genetic Hyping." *The Sciences,* March 12–15.

——. 2000. "It's Not 'All in the Genes.'" *Newsweek,* April 10.

Scarr, S. 1992. "Developmental Theories for the 1990s: Development and Individual Differences." *Child Development* 63 (1): 1516.

Scarr, S., and R. A. Weinberg. 1978. "The Influence of 'Family Background' on Intellectual Attainment." *American Sociological Review* 43: 674, 692.

Schmidt, M., and L. Moore. 1998. "Constructing a 'Good Catch,' Picking a Winner: The Development of Technosemen and the Deconstruction of the Monolithic Male." In *Cyborg Babies: From Techno-Sex to Techno-Tots,* edited by R. Davis-Floyd and J. Dumit, 21–39. New York: Routledge.

Schnieke, A., et al. 1997. "Human Factor IX Transgenic Sheep Produced by Transfer of Nuclei from Transfected Fetal Fibroblasts." *Science* 278: 2130–33.

Scott, J., and L. Fuller. *Genetics and the Social Behavior of the Dog.* Chicago: University of Chicago Press.

Sears, W., and M. White. 1999. *Nighttime Parenting.* Schaumburg, Ill.: La Leche League International.

Segal, N. 1999. "Entwined Lives: Twins and What They Tell Us about Human Behavior." New York: Penguin.

Siebert, C. 1996. "The Cuts That Go Deeper." *New York Times Magazine,* July 7, 20–26, 40–44.

Siegel, S. 1996. "Israeli, U.S. Scientists Find Risk Gene." *Jerusalem Post,* January 5.

Silver, L. 1997. *Remaking Eden: How Genetic Engineering and Cloning Will Transform the American Family.* New York: Avon Books.

Sinsheimer, R. 1989. "The Santa Cruz Workshop, May 1985." *Genomics* 5: 954–56.

Smaglik, P. 2000. "Chimeraplasty Potential." *Scientist* 14(1): 13.

Smith, C. 2000. "China Rushes to Adopt Genetically Modified Crops." *New York Times,* October 7.

Snouwaert, J., et al. 1992. "An Animal Model for Cystic Fibrosis Made by Gene Targeting." *Science* 257: 1083–88.

Spinney, L. 2000. "Switched On." *New Scientist,* November 4, 52–55.

Steen, G. 1996. *DNA and Destiny: Nature and Nurture in Human Behavior.* New York: Plenum, 161–83.

Sternberg, N., and D. Hamilton. 1981. "Bacteriophage Pi Site-Specific Recombination: I Recombination Between loxP Sites." *Journal of Molecular Biology* 150: 467–86.

Stevens, J., et al. 2001. "Haplotype Variation and Linkage Disequilibrium in 313 Human Genes." *Science* 293: 489–93.

Stock, G. 1993. *Metaman: Humans, Machines, and the Birth of a Global Superorganism.* New York: Simon & Schuster.

——. 2002. *The Book of Questions: In the Future.* In preparation.

Stock, G., and J. Campbell, eds. 2000. *Engineering the Human Germline: An Exploration of the Science and Ethics of Altering the Genes We Pass to Our Children.* New York: Oxford University Press.

Stourtin, E. "The Miracle Worker Battling Infertiiity and God's Wishes." *Daily Express,* May 4, 2000.

Strauss, E. 2001. "Growing Old Together." *Science* 292: 41–43.

Strober, W., and R. Ehrhardt. 1993. "Chronic Intestinal Inflammation: An Unexpected Outcome in Cytokine or T Cell Receptor Mutant Mice." *Cell* 75: 203–5.

Struewing, J., et al. 1997. "The Risk of Cancer Associated with Specific Mutations of BRCA1 and BRCA2 among Ashkenazi Jews." *New England Journal of Medicine* 336, (20), 1401–8.

Sunstein, C. 2001. "Slaughterhouse Jive." *New Republic,* January 29, 40–45.

Talbot, M. 2001. "A Desire to Duplicate." *New York Times Magazine,* February 4, 40–45.

Tambs, K., et al. 1989. "Genetic and Environmental Contributions to the Covariance Between Occupational Status, Educational Attainment, and IQ: A Study of Twins." *Behavior Genetics* 19: 202–22.

Tamkins, T. 1999. "Single Gene May Send You to Bed Early." *Reuters Health,* August 30.

Temple, L., et al. 2001. "Defining Disease in the Genomics Era." *Science* 293: 807–8.

Terman, L., and M. Oden. 1959. *Genetic Studies of Genius.* Vol. 5: *The Gifted Group at Mid-Life.* Stanford, Calif.: Stanford University Press.

Thornhill, R., and S. Gangestad. 1993. "Human Facial Beauty: Averageness, Symmetry, and Parasite Resistance." *Human Nature* 4, 237–69.

Trut, L. N. 1999. "Early Canid Domestication: The Farm Fox Experiment." *American Scientist* 87 (March).

Tsien, J. Z. 2000. "Building a Brainier Mouse." *Scientific American* 282: 62–68.

Turner, F. J. 1920. *The Frontier in American History.* New York: Henry Holt.

Vanfleteren, F., and A. De Vreese. 1995. "The Gerontogenes Age-1 and Daf-2 Determine Metabolic Rate Potential in Aging *Caenorhabditis elegans.*" *Journal of the Federation of the American Society of Experimental Biologists* 9 (13): 1355–61.

Vassaux, G. 1999. "New Cloning Tools for the Design of Better Transgenes." *Gene Therapy* 6: 307–8.

Venter, J., et al. 2001. "The Sequence of the Human Genome." *Science* 291: 1304–51.

Verrengia, J. 1999. "French Locate Seizure Gene." Associated Press newswire, October 29.

Viitanen, B., et al. 1997. "Heritability for Alzheimer's Disease: The Study of Dementia in Swedish Twins." *Journal of Gerontology: Medical Sciences* 52 (2): M117–125.

Vila, C., et al. 1997. "Multiple and Ancient Origins of the Domestic Dog." *Science* 76: 1687–89.

Vogel, G. 2001. "Human Cloning Plans Spark Talk of U.S. Ban." *Science* 292: 31.

Wade, N. 2000. "Reading the Book of Life." *New York Times,* June 27.

Wakayama, T., et al. 1998. "Full-Term Development of Mice from Enucleated Oocytes Injected with Cumulous Cell Nuclei." *Nature* 394: 369–74.

Warwick, K. 2000. "Cyborg 1.0." *Wired,* February, 145–51.

Wayne, R. 1993. "Molecular Evolution of the Dog Family." *Trends in Genetics* 9: 218–24.

Webb, S. 1907. *The Decline of the Birth Rate.* Fabian tract 131. London: The Fabian Society.

Weindruch, R., and R. Walford. 1988. *The Retardation of Aging and Disease by Dietary Restriction.* New York: Charles C. Thomas.

Weindruch, R., et al. 1986. "The Retardation of Aging by Dietary Restriction: Longevity, Cancer, Immunity, and Lifetime Energy Intake." *Journal of Nutrition* 116: 641.

Wertz, D. 1999. "Views of Chinese Medical Geneticists: How They Differ from 35 Other Nations." In *Chinese Scientists and Responsibility,* edited by Ole Döring, 141–60. Hamburg: Mitteilungen des Instituts für Asienkunde.

Wertz, R., and D. Wertz. 1989. *Lying-In: A History of Childbirth in America.* New Haven: Yale University Press.

Wetherell, J. L., et al. 1999. "History of Depression and Other Psychiatric Illness as Risk Factors for Alzheimer's Disease in a Twin Sample." *Alzheimer Disease and Associated Disorders* 13: 47–52.

Whitman, T. 2000. "DNA Detectives." *Industry Standard,* May 29, 166–88.

Whybrow, P. 1997. *A Mood Apart.* New York: HarperCollins.

Wilcox, A. 2000. "Quest for the Perfect Human Has Severe Flaws." *Los Angeles Times,* December 31, M5.

Will, G. 2001. "When Procreation Becomes Manufacture." *San Francisco Examiner,* January 22.

Willard, H. 1998. "Human Artificial Chromosomes Coming into Focus." *Nature Biotechnology* 16: 415.

———. 2000. "Perspectives: Artificial Chromosomes Coming to Life." *Science* 270: 1308–10.

Wilmut, I., et al. 1997. "Viable Offspring Derived from Fetal and Adult Mammalian Cells." *Nature* 385: 810–13.

Wilson, E. O. 1975. *Sociobiology: The New Synthesis.* Cambridge: Harvard University Press.

———. 1978. *On Human Nature.* Cambridge: Harvard University Press.

Winter, S. 2000. "Our Societal Obligation for Keeping Human Nature Untouched." *Engineering the Human Germline: An Exploration of the Science and Ethics of Altering the Genes We Pass to Our Children,* edited by G. Stock and J. Campbell, 113–16. New York: Oxford University Press.

Wivel, N., and L. Walters. 1993. "Germ-Line Gene Modification and Disease Prevention: Some Medical and Ethical Perspectives." *Science* 262: 537.

Woffendin, C., et al. 1996. "Expression of a Protective Gene Prolongs Survival of T Cells in Human Immunodeficiency Virus–Infected Patients." *Proceedings of the National Academy of Sciences* 93: 2889–94.

Wolfe, T. 1983. *The Right Stuff.* New York: Bantam Books.

——. 1999. "What Do a Jesuit Priest, a Canadian Communications Theorist, and Darwin II All Have in Common? Digibabble, Fairy Dust, Human Anthill." *Forbes,* October 4.

Wong, M. 2001. "Firm Launches Large-Scale Gene Project on Internet." *Los Angeles Times,* August 1, C1.

Wood, C., et al. 1997. "Cryopreservation of Ovarian Tissue: Potential 'Reproductive Insurance' for Women at Risk of Early Ovarian Failure." *Medical Journal of Australia* 166: 366–69.

Wooster, R., et al. 1994. "Localization of a Breast Cancer Susceptibility Gene, BRCA2, to Chromosome 13Q12–13." *Science* 265: 2088–90.

Worsley, F. 1931. *Endurance.* New York: W. W. Norton.

Wright, L. 1999. *Twins and What They Tell Us about Who We Are.* New York: John Wiley & Sons, 147.

Wright, R. 1994. *The Moral Animal: The New Science of Evolutionary Psychology.* New York: Pantheon Books.

Yamada, O., et al. 1994. "Intracellular Immunization of Human T Cells with a Hairpin Ribozyme Against Human Immunodeficiency Virus Type 1." *Gene Therapy* 1: 38–45.

Young, R., et al. 1999. "Site Specific Targeting of DNA Plasmids to Chromosome 19 Using AAV Cis and Tans Sequences. For: Gene Target Vector Protocols to Appear in Methods." In *Molecular Biology,* edited by E. B. Kmiec. Totowa, N.J.: Humana Press.

Young, Y., et al. 1999. "Increased Affiliative Response to Vasopressin in Mice Expressing the V1a Receptor from a Monogamous Vole." *Nature* 400: 766–68.

Yunis, J., et al. 1980. "The Striking Resemblance of High Resolution G-Banded Chromosomes of Man and Chimpanzee." *Science* 208: 1145–48.

Zeng, Y., et al. 1993. "Causes and Implications of the Recent Increase in the Reported Sex Ratio at Birth in China." *Population and Development Review* 19: 2.

Zivin, J. 2000. "Understanding Clinical Trials." *Scientific American,* April, 69–75.

Index